George Crispe Whiteley

The Law relating to Weights, Measures and Weighing Machines

George Crispe Whiteley

The Law relating to Weights, Measures and Weighing Machines

ISBN/EAN: 9783337339487

Printed in Europe, USA, Canada, Australia, Japan

Cover: Foto ©berggeist007 / pixelio.de

More available books at **www.hansebooks.com**

THE LAW

RELATING TO

WEIGHTS, MEASURES, AND WEIGHING MACHINES.

BY

GEORGE CRISPE WHITELEY,

M.A. CANTAB., BARRISTER-AT-LAW.

"*Be just, and fear not.*"
HEN. VIII., Act iii., sc. 2.

LONDON: KNIGHT AND CO., FLEET STREET.
EDINBURGH: BELL AND BRADFUTE, BANK STREET.
DUBLIN: HODGES, FOSTER, AND FIGGIS, GRAFTON STREET.
1879.

LONDON:
PRINTED BY WILLIAM CLOWES AND SONS,
STAMFORD STREET AND CHARING CROSS.

PREFACE.

THE law relating to weights, measures, and weighing machines is now, for the most part, contained in the "Weights and Measures Act, 1878," which came into operation upon the first day of January in the present year. The arrangement of this book follows closely the order of the Act itself, and notes have been added to the different sections which, it is hoped, will be of use both to those who have to obey the law, and to those who have to carry the provisions of the statute into effect. In preparing these notes, the Memorandum prefixed to the Act when passing through Parliament has been freely used; and much help has been found in the twelve Annual Reports of the "Warden of the Standards," issued during the years 1866–78. To make the book complete, the enactments imposing the obligation to keep weights in a mill, imposing penalties for false weights under the Excise Acts, providing for scales and weights at markets and fairs, and relating to weights and measures used in mines, have been given, and will be found in Chapter X.; and with a similar view, a chapter has been added giving the provisions of the law with regard to the sale of coals and bread, so far as they relate to weights and measures. My acknowledgments are due, and are here given, to a member of the Scotch Bar for revising the portion of Chapter XI. which deals with Scotland; and I have to thank Mr. Strugnell and Mr. Webb, the able Inspectors of the Newington Division of the County of Surrey, for many valuable practical suggestions.

G. C. W.

The Temple,
 March 1879.

CONTENTS.

CHAP.		PAGE
	INDEX TO STATUTE	ix
	TABLE OF CASES	xi
	INTRODUCTION	xiii
I.	IMPERIAL MEASURES OF LENGTH, WEIGHT, AND CAPACITY	1
II.	STANDARDS OF MEASURE AND WEIGHT	7
	(1.) Imperial Standards	7
	(2.) Parliamentary Copies of Imperial Standards	8
	(3.) Secondary or Board of Trade Standards	9
	(4.) Local Standards	11
III.	METRIC WEIGHTS AND MEASURES	13
IV.	ADMINISTRATION	15
	(1.) Central—The Board of Trade	15
	(2.) Local—The Local Authority	19
V.	THE USE OF WEIGHTS AND MEASURES	29
VI.	UNJUST WEIGHTS, MEASURES, AND WEIGHING MACHINES	41
VII.	STAMPING AND VERIFICATION OF WEIGHTS AND MEASURES	48
VIII.	INSPECTORS AND THEIR DUTIES	55
IX.	LEGAL PROCEEDINGS	64
	(1.) Summary Proceedings	64
	(2.) Proceedings upon Appeal	73
	(3.) Actions against Inspectors and Others	78
X.	MISCELLANEOUS	80
	(1.) Legal Provisions	80
	(2.) Savings and Definitions	80
	(3.) Repeal	84
	(4.) Miscellaneous provisions	86

CHAP.		PAGE
XI. SCOTLAND AND IRELAND	. . . .	93
(1.) Scotland	. . . .	93
(2.) Ireland	. . . .	101
XII. THE SALE OF COALS AND BREAD	. . . . .	117
(1.) The Sale of Coals	. . . .	117
(2.) The Sale of Bread	. . . .	129
CONTENTS OF APPENDIX	. . . .	135
APPENDIX	. . . .	137
INDEX	. . . .	201

INDEX TO STATUTE.

The Weights and Measures Act, 1878.
41 & 42 Vict. c. 49.

Sect.	Page	Sect.	Page
1	..	37	17
2	..	38	13
3	29	39	19
4	7	40	19
5	8	41	23
6	7	42	24
7	8	43	55
8	9	44	57
9	11	45	54
10	1	46	49
11	1	47	57
12	2	48	60
13	2	49	63
14	3	50	20
15	4	51	20
16	5	52	21
17	5	53	25
18	13	54	26
19	31	55	26
20	32	56	64
21	13	57	70
22	33	58	71
23	39	59	71
24	32	60	73
25	41	61	78
26	41	62	86
27	42	63	80
28	48	64	80
29	48	65	80
30	48	66	80
31	52	67	81
32	54	68	81
33	15	69	81
34	16	70	83
35	16	71	93
36	16	72	94

Index to Statute.

SECT.	PAGE	SECT.	PAGE
73	95	81	105
74	97	82	107
75	99	83	107
76	101	84	108
77	102	85	109
78	102	86	84
79	104		
80	105	Schedules	87, 88, 113, 117

TABLE OF CASES.

	PAGE
Aërated Bread Company v. Gregg	131
Blandford v. Morrison	121
Booth v. Shadgett	43
Carr v. Stringer	47
Collins v. Hopwood	128
Cundell v. Dawson	120
Davie v. Robertson	100
Duly v. Sharood	22, 56
Frend v. Butterfield	125
Goody v. Penny	117
G. W. R. v. Bailie	46
Griffiths v. Place	61
Henry v. McEwan	100
Hill v. Browning	131
Hockin v. Cooke	36
Hughes v. Humphreys	33
Hutchins v. Reeves	61
Jones v. Giles	36
Jones v. Huxtable	130
Kershaw v. Johnson	61, 62
Little v. Poole	120
L. & N. W. R. v. Richards	46
Megarry v. McCullagh	102
Meredith v. Holman	120, 125
Miller v. Muir	100
Mitton v. Troke	130
Moore v. Wicker	82

	PAGE
Owens v. Denton	33
Painter v. Seers	47
Paterson v. Robertson	100
R. v. Aulton	51
R. v. Dunnage	41
R. v. Eagleton	41
R. v. Jarvis	83
R. v. Kennett	131
R. v. Kingsley	133
R. v. Receiver of Hull	22, 56
R. v. Saunders	131
R. v. Skelton	59
R. v. Wheatley	41
R. v. William Wood	130
R. v. Young	41
Robertson v. Good	100
Robertson v. Hart	100
Robinson v. Cliff	133
St. Cross v. Howard	36
Shoreditch Guardians v. Franklin	118
Smith v. Cartwright	117
Starr v. Stringer	51
Starr v. Trinder	51
Stevenson v. Sheckle	86
Thomas v. Stevenson	44, 62, 79
Tyson v. Thomas	33
Washington v. Young	51
Willcock v. Windsor	82
Williams v. Deggan	130
Withall v. Francis	47
Wray v. Reynolds	46

INTRODUCTION.

The appointment of standard weights and measures, which, as the criteria of value, ought to be as certain and uniform as possible, is delegated by the law of this country to the King, as in Normandy it was vested in the Duke. However, although this power is said to be inherent in the Crown, it has been exercised from the earliest times by the Legislature.*

It will be unnecessary here to trace the history of the various enactments which have been passed from time to time to regulate weights and measures in this country. Modern legislation may be said to have begun in 1795, 1797, when Acts were passed providing for "the more effectual prevention of the use of defective weights, and of false and unequal balances." These were followed in 1815 by an Act which dealt in a similar way with measures; and in 1824, 1825, provisions were made for "ascertaining and establishing uniformity of weights and measures." These five statutes, although not expressly repealed until 1878, were for the most part superseded by

---

* 1 Bla. Comm. 274.

the Act of 1835 (5 & 6 Will. IV. c. 63), which, as amended by the Act of 1859, has been the principal enactment regulating the use of weights, measures, and weighing machines up to the passing of the "Weights and Measures Act, 1878."

The object of this last Act was to "consolidate (for the purpose of statute law revision), and not to amend the law." It is described as "a simple reproduction of the existing statute law in a compact and consistent form, without amending it by the introduction of more effective procedure, or otherwise substantially altering it." The general principles respecting the obligation to use imperial weights and measures, the prohibition of unjust weights and measures, the verification and inspection of weights and measures, and the partial legalisation of metric weights and measures, are unchanged.

The Memorandum prefixed to the Act during its passing through Parliament, which has been freely used in the present book, describes certain amendments of detail which, under the circumstances, were inevitable. These amendments are classed under three heads, according as they arise:

1. From the mere act of consolidation;
2. From change of circumstances produced by the alteration of the general law; by a practice which has grown up, and by facts which have occurred during the lapse of time, and which must be recognised; or

3. From the desirability of removing doubts and correcting mistakes.

Thus, under the first head, we have a uniform system laid down for the whole of the United Kingdom, in the place of different provisions for England and Scotland, on the one hand, and for Ireland on the other; and the omission of all enactments which relate to formal matters, and where the operation is spent. Under the second head may be classed the amendments made where an enactment has, by lapse of time or by the alteration of the general law, or by reason of its having ceased, become obsolete; for example, the substitution of the procedure in Jervis's Act for the old procedure for recovering penalties. Under the third head comes the removal of doubts which have arisen on the wording of the former Act, and which it is undesirable to leave. These doubts have been solved in accordance with the existing practice. There are also amendments to supply obviously accidental omissions, and for the correction of obvious mistakes.

In addition to the amendments coming under one of the above heads, there are others, of detail and not of principle, which it is important to note:

1. In several sections the maximum penalty is increased in the case of a second offence;
2. The use of material weights above fifty-six pounds, until the Queen in Council legalises a new standard, is expressly prohibited, and

the exception of those weights and of wooden and earthenware measures from being verified is removed;

3. The local verification of standards is facilitated;

4. Power is given to local authorities to combine for the purpose of standards and inspection;

5. The express power to verify and stamp measures, made partly of metal and partly of glass, which is now confined to measures for exciseable liquids, is extended to measures for all liquids;

6. The pecuniary penalty for the sale by heaped measure, over and above the avoidance of the contract, is removed;

7. The prohibition on the use of authorised weights, &c., and the power to inspect weights, &c., and to enter premises for that inspection, are made more uniform and complete;

8. Certain denominations of Board of Trade standards which are scarcely used in practice have been omitted from the schedule, and consequently will cease to be standards;

9. The power of appointing examiners of weights and measures is omitted;

10. Power is given to the Queen in Council on local application to alter and add to the inspectors' fees.

Introduction.

These amendments have all been pointed out in the notes which follow the sections in which they occur; and reports upon the cases which have been decided upon the construction of the former Acts have been inserted for the purpose of assisting in the correct interpretation of the provisions of the present statute.

THE LAW

RELATING TO

WEIGHTS, MEASURES, AND WEIGHING MACHINES.

CHAPTER I.

IMPERIAL MEASURES OF LENGTH, WEIGHT, AND CAPACITY.

The straight line or distance between the centres of the two gold plugs or pins in the bronze bar declared to be the imperial standard for determining the imperial standard yard measured when the bar is at the temperature of sixty-two degrees of Fahrenheit's thermometer, and when it is supported on bronze rollers placed under it in such manner as best to avoid flexure of the bar, and to facilitate its free expansion and contraction from variations of temperature, is the legal standard measure of length, and is called the imperial standard yard, and is the only unit or standard measure of extension from which all other measures of extension, whether linear, superficial, or solid, are ascertained. *Imperial standard yard. 41 & 42 Vict. c. 49, s. 10.*

One third part of the imperial standard yard is a foot, and the twelfth part of such foot is an inch, and the rod, pole, or perch in length contains five such yards and a half, and the chain contains twenty-two such yards, the furlong two hundred and twenty such yards, and the mile one thousand seven hundred and sixty such yards. *Linear measures derived from imperial standard yard. 41 & 42 Vict. c. 49, s. 11*

Superficial measures derived from the imperial standard yard.
41 & 42 Vict. c. 49, s. 12.

The rood of land contains one thousand two hundred and ten square yards according to the imperial standard yard, and the acre of land contains four thousand eight hundred and forty such square yards, being one hundred and sixty square rods, poles, or perches.

The imperial standard of measure referred to is the bronze bar which is declared to be the imperial standard of measure in section 4 of the Weights and Measures Act, 1878. It is fully described in the first schedule of the said Act, which will be found in the *Appendix*.

These definitions of the imperial measures of length are repetitions of those to be found in the Acts of 1824 and 1855, both of which are now repealed, with the addition of the description of the manner in which the bar forming the standard yard is to be supported when used for ascertaining the legal standard measure.

The chain shall contain twenty-two yards. This also is an addition, for although the Board of Trade standard for a chain has been legalised by Order in Council, and is thus a recognised measure, it is here made an imperial measure "ascertained by the Weights and Measures Act," the importance of which will be noticed when the secondary standards of measure, and the use of imperial measures are considered. See *post*. The link, the hundredth part of a chain, a sub-division extensively used by surveyors is thus made a legal measure, being an aliquot part of an imperial measure.

This imperial yard is the unit of length. By the Act passed in 1824 (5 Geo. IV. c. 74) this yard was described (sect. 3) to be the length of the pendulum, vibrating seconds of mean time in the latitude of London in a vacuum at the level of the sea, in the proportion of 36 to 39·1393. This declaration was made for the purpose of replacing the standard yard, if lost, destroyed, defaced, or otherwise injured, by reference to some "invariable natural standard." When, however, it became necessary to restore the standards, after the burning down of the Houses of Parliament, the foregoing section was repealed, and the present standard was constructed from a comparison of copies that had been carefully made of the old standard. See 18 & 19 Vict. c. 72, now repealed.

Imperial standard pound.
41 & 42 Vict. c. 49, s. 13.

The weight *in vacuo* of the platinum weight declared to be the imperial standard for determining the imperial standard pound, is the legal standard measure of weight, and of measure having reference to weight, and is called the imperial standard pound, and is the only unit or standard measure of weight from which all other weights and all measures having reference to weight are ascertained.

One-sixteenth part of the imperial standard pound is an ounce, and one-sixteenth part of such ounce is a dram, and one seven-thousandth part of the imperial standard pound is a grain.

A stone consists of fourteen imperial standard pounds, and a hundredweight consists of eight such stones, and a ton consists of twenty such hundredweights.

Four hundred and eighty grains are an ounce troy.

All the foregoing weights except the ounce troy are deemed to be avoirdupois weights.

Imperial weights derived from imperial standard pound. 41 & 42 Vict. c. 49, s. 14.

The imperial standard of weight referred to is the platinum weight which is declared to be the imperial standard of weight in section 4 of the Weights and Measures Act, and is fully described in the first schedule of that Act, which will be found in the *Appendix.*

The unit of weight is the pound avoirdupois, which is now the imperial standard pound described as above. Originally the standard pound was a brass weight of one pound troy weight made in 1758, and from this unit were derived the troy ounce, pennyweight, and grain, and from the grain were computed the pound, ounce, and dram avoirdupois (5 Geo. IV., c. 74, s. 4). Provision was made in the same Act (s. 5) for replacing the same standard troy pound, if lost, destroyed, defaced, or otherwise injured, by reference to some invariable natural standard, and it was declared that a cubic inch of distilled water, weighed in air by brass weights at 62° F., the barometer being at 30 inches, is equal to 252·458 grains; that 7000 such grains were a pound avoirdupois, and 5760 a pound troy. However, as in the case of the standard yard, when the pound troy and the pound avoirdupois were lost by the destruction of the Houses of Parliament, the foregoing sections were repealed, and by sect. 3 of the 18 & 19 Vict. c. 72, the present standard pound avoirdupois was declared to be the imperial standard pound, and this was to be deemed the only standard measure of weight from which all other weights were to be derived, computed, and ascertained, and one equal seven-thousandth part of such pound was to be a grain, and 5760 such grains were to be a pound troy. In the meantime, in 1835, by sect. 11 of 5 & 6 Will. IV. c. 63, the stone weight was declared to consist of 14 standard pounds avoirdupois; the hundredweight to consist of 8 such stones, and the ton to consist of 20 such hundredweights.

These enactments are also now repealed, and the sections given above take their place. The troy pound, and troy pennyweight are omitted, the sales by troy weight being limited, both in the Sales of Bullion Act (16 & 17 Vict. c. 29), now repealed, and in the Weights and Measures Act, 1878, to the ounce troy, or any multiple or decimal part of such ounce.

In vacuo.—The provision that the weight should be weighed *in vacuo* was not in the former statutes.

No mention is made in these sections of apothecaries' weight, although the sale by retail of drugs by this weight is expressly permitted. The apothecaries' weight, however, is only a sub-division of troy weight, there being 20 grains to the scruple, 3 scruples to the dram, and 8 drams to the ounce, which is the same as the ounce troy. This weight, however, is chiefly used for dispensing purposes, drugs being now sold almost invariably by avoirdupois weight.

<small>Imperial measures of capacity. 41 & 42 Vict. c. 49, s. 15.</small>

The unit or standard measure of capacity from which all other measures of capacity, as well for liquids as for dry goods, are derived, is the gallon containing ten imperial standard pounds weight of distilled water weighed in air against brass weights, with the water and the air at the temperature of sixty-two degrees of Fahrenheit's thermometer, and with the barometer at thirty inches.

The quart is one-fourth part of the gallon, and the pint is one-eighth part of the gallon.

Two gallons are a peck, and eight gallons are a bushel, and eight such bushels are a quarter, and thirty-six such bushels are a chaldron.

The unit of capacity is the gallon. The imperial standard gallon is defined in the above section, a definition taken for the most part from the Act of 1824 (5 Geo. IV. c. 74, s. 6). That Act also provided that a measure should be made of brass to represent the imperial standard gallon, but did not provide for its restoration if injured. That measure was made, and is included in the list of Board of Trade standards, which are to be found in the second schedule of the Weights and Measures Act, 1878, given in the *Appendix*. As the capacity of the gallon is defined by reference to the standard pound, there was no necessity for an imperial standard gallon. By making, therefore, the existing measure a Board of Trade standard, it can be more easily replaced if lost, defaced, or injured. See *post*.

The provisions that the water should be weighed against brass weights, and that the air as well as the water should be at the temperature of 62°, are not found in the previous enactments.

In the Act of 1824 the "quarter" was limited to "corn, or other dry goods, not measured by heaped measure." As, however, heaped measure was afterwards (1835) abolished, this limitation is now omitted. No mention also is made of a "sack," which is a most important omission, more particularly with reference to the sale of coke which is still sold by measure. Hitherto the only legal definition of a "sack" was contained in section 8 of 5 Geo. IV. c. 74, which enacted that "three bushels should be a sack." Twelve of these sacks were a chaldron, which thus contained thirty-six bushels. But these bushels were "heaped bushels," and all heaped measure was abolished in 1835. It therefore became a nice point of law as to whether the sack of coke should still contain three heaped bushels or three stricken bushels, and the amount of a chaldron of coke varied in consequence. In practice the large London gas companies have taken the capacity of the sack and chaldron to be derived from the heaped bushel, and no change was made by them since, or in consequence of the passing of the Act of 1835, and owing to the competition amongst the several companies even fuller measure than this is continually given. But the retail dealers appear generally to have sold by the stricken measure, and as the sack in the one case contained 8446·4661 cubic inches, and in the other 6654·576 cubic inches, the difference was considerable. Now, however, the "sack" is omitted altogether, and it becomes an unrecognised measure. The chaldron is defined to be thirty-six bushels, each bushel containing eight imperial gallons.

A bushel for the sale of any of the following articles, namely, lime, fish, potatoes, fruit, or any other goods and things which before the ninth day of September, one thousand eight hundred and thirty-five, were commonly sold by heaped measure, shall be a hollow cylinder having a plane base, the internal diameter of which shall be double the internal depth, and every measure used for the sale of any of the above-mentioned articles which is a multiple of a bushel, or is a half bushel or a peck, shall be made of the same shape and proportion as the above-mentioned bushel. *Measure of capacity for goods formerly sold by heaped measure. 41 & 42 Vict. c. 49, s. 16.*

In using an imperial measure of capacity, the same shall not be heaped, but either shall be stricken with a round stick or roller, straight and of the same diameter from end to end, or if the article sold cannot from its size or shape be conveniently stricken, shall be filled in all parts as nearly to the level of the brim as the size and shape of the article will admit. *Measure of capacity, when used, to be stricken or filled up, 41 & 42 Vict. c. 49, s. 17.*

"Heaped measure" was abolished by the 5 & 6 Will. IV. c. 63, s. 7, which came into operation upon the date mentioned in the above section, September 9th, 1835. This measure originally applied to coals, culm, lime, fish, potatoes, and fruit, and other articles which, from their size and shape, were not capable of being stricken. For these articles the standard measure of capacity was a bushel, containing eighty pounds avoirdupois of water, the same being made round, with a plain and even bottom, and being 19½ inches from outside to inside (5 Geo. IV. c. 74, s. 7). In using this bushel the articles mentioned were to be "heaped up" in the manner therein described; but for all other articles "stricken measure" was to be used. In the next year, 1825, the figure of other measures used in selling by heaped measure was fixed and determined. They were to be cylindrical, and their diameter was to be at least double their depth (6 Geo. IV. c. 12. s. 2). In 1835, as before mentioned, heaped measure was abolished, and all "coals, culm, slack, and cannel" were to be sold by weight and not by measure. There still remained the lime, fish, potatoes, fruit, and other articles which "from their size and shape were incapable of being stricken, and from their nature and quality might not conveniently be sold by weight." For these articles it was enacted that they might be sold by weight; but if sold by measure, they were to be sold in similar measures as heretofore, and the measures were to be "filled in all parts as nearly to the level of the brim as the size and shape of the articles would admit." In the place of these enactments we have the two sections given above. They apply to the same articles that were formerly sold by heaped measure. There seems to be no obligation to use the bushel, which, if used for these articles, is to be of the shape described. The size of the diameter of the bushel (19½ inches) is omitted, and the measures smaller than the bushel are limited to the half bushel and peck, the practical result of which is that no shape is prescribed for the gallon or any less measure in commercial use.

In the Twelfth Annual Report of the Weights and Measures Department of the Board of Trade (1877-78) an account is given of experiments which have recently been made upon the different modes of "striking" a measure, or of bringing the contents of a measure into the same horizontal plane as that of the brim of the measure. The true weight of a standard bushel measure of a certain sort of corn was found to be 57 lb. 2 oz.; the weight of the measure when struck with the ordinary flat strike was found to be 57 lb. 3 oz., and when struck with the ordinary round stick or roller it was found to be 57 lb. 9 oz. If when struck with the round stick the measure was then shaken, the weight became 62 lb. 15 oz.

CHAPTER II.

STANDARDS OF MEASURE AND WEIGHT.

(1.) *Imperial Standards.*

The bronze bar and the platinum weight, more particularly described in the first part of the First Schedule to the Weights and Measures Act, 1878, and at the passing of that Act deposited in the Standards Department of the Board of Trade in the custody of the Warden of the Standards, are the imperial standards of measure and weight, and the said bronze bar is the imperial standard for determining the imperial standard yard for the United Kingdom, and the said platinum weight is the imperial standard for determining the imperial standard pound for the United Kingdom. <small>Imperial standards of measure and weight. 41 & 42 Vict. c. 49, s. 4.</small>

If at any time either of the imperial standards of measure and weight is lost or in any manner destroyed, defaced, or otherwise injured, the Board of Trade may cause the same to be restored by reference to or adoption of any of the parliamentary copies of that standard, or of such of them as may remain available for that purpose. <small>Restoration of imperial standards. 41 & 42 Vict. c. 49, s. 6.</small>

<small>There are four kinds of standards: 1. The imperial standards; 2, the Parliamentary copies of the imperial standards; 3, the secondary, or Board of Trade standards; and 4, the local standards.

The imperial standards mentioned in the above sections have already been incidentally mentioned in dealing with the imperial measures of length and weight (Chap. I.). Originally copies of the old standards, which were destroyed in the old Houses of Parliament, they were, after careful examination, declared to be the imperial standards by sections 2 & 3 of the 18 & 19 Vict. c 72. They were deposited in the office of the Exchequer, and in 1866 were duly transferred, under the provisions of the 29 and 30 Vict.</small>

c. 82, to the possession of the Board of Trade, in whose custody they still remain. See Administration, *post*. The imperial standards are fully described in the first schedule of the Weights and Measures Act, 1878, which will be found in the *Appendix*.

The provision for restoring or replacing the imperial standards, if lost, destroyed, or defaced, by referring to or adopting any of the parliamentary copies, is a re-enactment of a similar clause now repealed in the Act of 1855 (18 & 19 Vict. c. 72, s. 7).

(2.) *Parliamentary Copies of Imperial Standards.*

Parliamentary copies of imperial standards.
41 & 42 Vict. c. 49, s. 5.

The four copies of the imperial standards of measure and weight, described in the second part of the First Schedule to the Weights and Measures Act, 1878, and deposited as therein mentioned, shall be deemed to be parliamentary copies of the said imperial standards.

The Board of Trade shall, as soon as may be after the commencement of the said Act, cause an accurate copy of the imperial standard of measure, and an accurate copy of the imperial standard of weight to be made of the same form and material as the said standards, and it shall be lawful for Her Majesty in Council, on the representation of the Board of Trade, to approve the copies so made, and the copies when so approved shall be of the same effect as the said parliamentary copies, and are in the said Act included under the name parliamentary copies of the imperial standards of measure and weight.

Restoration of Parliamentary copies.
41 & 42 Vict. c. 49, s. 7.

If at any time any of the parliamentary copies of either of the imperial standards is lost or in any manner destroyed, defaced, or otherwise injured, the Board of Trade may cause the same to be restored by reference either to the corresponding imperial standard, or to one of the other parliamentary copies of that standard.

There are to be five parliamentary copies of the imperial standards. Four of these are deposited at the Royal Mint, with the Royal Society of London, at the Royal Observatory of Greenwich, and at the Houses of Parliament respectively. These four copies were made at the same time as the imperial standards. The Board of Trade are required to make a fifth copy, which is to have the same effect as the other copies, and which will be for the use of the Board of Trade in verifying the secondary standards, and so diminish the use of the imperial standards, and thereby save their gradual depreciation. The parliamentary copies of the imperial standards, which have been deposited with the authorities at the Mint, Royal Society, and Royal Observatory, and immured in the new palace at Westminster, are to continue to be so deposited, and the copies, which are to be made in accordance with the provisions of the Weights and Measures Act, 1878, are to be in the custody of the Board of Trade. Provision is made also in the Weights and Measures Act, 1878, for the periodical verification of the parliamentary copies with the imperial standards and with one another. See Administration, *post*.

The power to restore any of the parliamentary copies of either of the imperial standards, by referring either to the imperial standards themselves or to any of the other copies, appears for the first time in the Act of 1878. The parliamentary copies are fully described in the second part of the first schedule of the Weights and Measures Act of 1878, which will be found in the *Appendix*.

(3.) *Secondary or Board of Trade Standards.*

The secondary standards of measure and weight which, having been derived from the imperial standards, are, upon the first day of January 1879, in use under the direction of the Board of Trade, and are mentioned in the Second Schedule to the Weights and Measures Act, 1878, and no others (save as hereinafter mentioned) shall be secondary standards of measure and weight, and shall be called Board of Trade standards.

<small>Secondary (Board of Trade) standards of measure and weight. 41 & 42 Vict. c. 49, s. 8.</small>

If at any time any of such standards is lost or in any manner destroyed, defaced, or otherwise injured, the Board of Trade may cause the same to be restored by reference either to one of the imperial standards or to one of the parliamentary copies of those standards.

The Board of Trade shall from time to time cause such new denominations of standards, being either equivalent to, or multiples, or aliquot parts of the imperial weights and measures ascertained by this Act, or being equivalent to or multiples of each coin of the realm for the time being, as appear to them to be required, in addition to those mentioned in the Second Schedule to the Weights and Measures Act, 1878, to be made and duly verified, and those new denominations of standards when approved by Her Majesty in Council shall be Board of Trade standards in like manner as if they were mentioned in the said schedule.

It shall be lawful for Her Majesty by Order in Council to declare that a Board of Trade standard for the time being of any denomination, whether mentioned in the said schedule or approved by Order in Council, shall cease to be such a standard.

Such standards of the Board of Trade as are equivalent to or multiples of any coin of the realm for the time being shall be standard weights for determining the justness of the weight of and for weighing such coin.

The second schedule to the Weights and Measures Act, 1878, will be found in the *Appendix*. It contains a list of all the Board of Trade standards which " are in use," at the commencement of the Act, that is to say upon the first day of January 1879. These standards are either those which were recognised as legal by the standards of Weights, Measures, and Coinage Act, 1866 (29 & 30 Vict. c. 82, s. 3), or have been legalised by an Order in Council under section 6 of the said Act. Some of these standards, which were found in practice to be of but little use, have been omitted from the present schedule, and will consequently cease to be Board of Trade standards, unless again legalised by Order in Council.

By the 24th section of the Weights and Measures Act, 1878, every person who " uses or has in his possession for use for trade " a weight or measure which is not of the denomination of some Board of Trade standard is liable to a penalty. So that it is of the highest importance to note the list of Board of Trade standards as given in the schedule, the possession of any other kind of weight or measure being illegal.

In consequence of representations made by chemists, there have been added to the schedule the standards of fluid ounces, fluid drams, and minims. These measures are in very extensive use, and are recognised in the Pharmacopœia.

The "bottle" and "half-bottle" measures, which by Order in Council dated March 24, 1871, were declared to be " legal secondary standards of capacity," and were defined to be equal to one-sixth and one-twelfth of a gallon respectively, are omitted from the present schedule, so that the purchasers of wine are now thrown back upon the old system under which they must either insist upon receiving imperial measure, or be content to take their wines in an unknown quantity delivered to them in vessels which are "not represented" as containing any amount of imperial measure." This question will be again dealt with in a future chapter upon the "Use of Imperial Measures." It is very fully considered in the fifth annual report of the Warden of the Standards, 1870-71.

There has been a movement throughout the country in favour of establishing a standard of 100 lbs. for the sale of grain, flour, meal, &c., and this proposition having received a considerable amount of support the material standard weight of the "cental" of 100 lbs. has now been legalised by the Board of Trade, in accordance with the powers given to them in the above section of the Weights and Measures Act, 1878. See Order in Council, dated February 4th, 1879, in the *Appendix*.

These secondary standards are to be in the custody of the Board of Trade, and once at least in every five years are to be compared with the parliamentary copies of the imperial standards, and with each other, and, if necessary, are to be thereupon adjusted or renewed. See Administration, *post*.

(4.) *Local Standards.*

The standards of measure and weight which are upon the first day of January 1879, legally in use by inspectors of weights and measures for the purpose of verification or inspection, and all copies of the Board of Trade standards which after the date aforesaid are compared with those standards and verified by the Board of Trade for the purpose of being used by inspectors of weights and measures under the Weights and Measures Act, 1878, as standards for the verification or inspection of weights and measures, shall be called local standards.

Local standards of measure and weight. 41 & 42 Vict. c. 49, s. 9.

In the Act of 1835 (5 & 6 Will. IV. c. 63, s. 4), it was enacted that the local standards need not be models or copies in shape or form of the Board of Trade standards. As, however, in practice these local standards are invariably copies in shape and form of the Board of Trade standards, this provision is omitted in the present section. In practice also the Board of Trade standards are alone used for verifying the local standards. The Board of Trade is to compare and verify, and re-verify, all copies of any of these standards which are submitted for the purpose by any local authority; and the local authority is bound to provide such local standards of measure and weight as they may deem requisite for the purpose of the comparison by way of verification or inspection of all weights and measures in use within their jurisdiction (41 & 42 Vict. c. 49, ss. 37, 40). Provision is also made for the re-verification of local standards, for their production by the person having the custody of them upon reasonable notice, and for defraying the expense of providing them. See 41 & 42 Vict. c. 49, ss. 41, 42, and 51; and Administration, *post*.

CHAPTER III.

METRIC WEIGHTS AND MEASURES.

The table in the Third Schedule to the Weights and Measures Act, 1878, shall be deemed to set forth the equivalents of imperial weights and measures, and of the weights and measures therein expressed, in terms of the metric system, and such table may be lawfully used for computing and expressing, in weights and measures, weights and measures of the metric system. Equivalents of metric weights and measures in terms of imperial weights and measures. 41 & 42 Vict. c. 49, s. 18.

A contract or dealing shall not be invalid or open to objection on the ground that the weights or measures expressed or referred to therein are weights or measures of the metric system, or on the ground that decimal subdivisions of imperial weights and measures, whether metric or otherwise, are used in such contract or dealing. Exception for contract, &c., in metric weights and measures. 41 & 42 Vict. c. 49, s. 21.

The Board of Trade may, if they think fit, cause to be compared with the metric standards in their custody and verified all metric weights and measures which are submitted to them for the purpose, and are of such shape and construction as may be from time to time in that behalf directed by the Board of Trade, and which the Board of Trade are satisfied are intended to be used for the purpose of science or of manufacture, or for any lawful purpose not being for the purpose of trade within the meaning of the Weights and Measures Act, 1878. Power of Board of Trade to verify metric weights and measures. 41 & 42 Vict. c. 49, s. 38.

The table of metric equivalents will be found in the third schedule of the Weights and Measures Act, 1878, which is given in the *Appendix*.

Not being for the purpose of trade within the meaning of the Weights and Measures Act, 1878.—This refers to the definition of "trade" given in the 19th section of the said Act:—"Every contract, bargain, sale, or dealing made or had in the United Kingdom for any work, goods, wares, or merchandise or other thing which has been or is to be done, sold, delivered, carried, or agreed for by weight or measure, and all tolls and duties charged or collected, according to weight or measure."

In 1874 was passed an "Act to render permissive the use of the metric system of weights and measures" (27 & 28 Vict. c. 117). The preamble stated that "it was expedient to legalise the use of the metric system of weights and measures for the promotion and extension of our internal as well as our foreign trade, and for the advancement of science," and the second section enacted that "no contract or dealing should be deemed to be invalid or open to objection on the ground that it was made in terms of the metric system." This Act is now repealed, and the sections given above take its place. It will be observed that although contracts and dealings are permitted to be expressed or referred to in terms of the metric system, and metric weights and measures are lawful "if used for the purpose of science or of manufacture," they cannot be used for trading purposes, and if so used, the person using the same would be liable to a penalty under the 24th section of the Weights and Measures Act, 1878, for using or having in possession for use, *for trade*, an unauthorised weight or measure, and under the 48th section of the same Act an inspector would have authority to seize any weight or measure so used *for trade*, which by the 24th section above-mentioned is liable to be forfeited.

The French are responsible for the introduction of the metric system. At one time there existed in France the same want of uniformity in weights and measures as exists at the present time in England. Soon after the Revolution of 1789 a Commission was appointed, consisting of Lagrange, Laplace, and others, to prepare a new system of weights and measures. The system so prepared was established by the French Legislature in 1801, and now prevails throughout the whole of France. It is known as the "Metric System," and it has been introduced in Holland, Belgium, and Switzerland. It is also used, more or less, in Italy, Germany, and the United States. The unit of length is the *Metre*, which is also the fundamental unit, because from it every other unit of weight or measure is derived, and thus we get the name *metric system*. A metre is defined to be "the ten-thousandth part of the distance from the pole to the equator, measured along the surface of the ocean." In forming the multiples and sub-multiples, the decimal system is followed exclusively; the Greek prefixes to any unit denoting multiples and the Latin submultiples. Thus ten metres is a deca-metre, the tenth part of a metre is a decimetre; one hundred metres is a hecto-metre; the one-hundredth part of a metre is a centimetre, and so on. The unit of surface is the *square metre*; of volume, the *cubic metre*; of capacity, the *litre*; of weight, the *gramme*; and of money, the *franc*.

CHAPTER IV.

ADMINISTRATION.

(1.) *Central.—The Board of Trade.*

The Board of Trade shall have all such powers and perform all such duties relative to standards of measure and weight, and to weights and measures, as are by any Act or otherwise vested in or imposed on the Treasury, or the Comptroller-General of the Exchequer, or the Warden of the Standards; and all things done by the Board of Trade, or any of their officers, or at their office, in relation to standards of weights and measures in pursuance of the Weights and Measures Act, 1878, shall be as valid, and have the like effect and consequences, as if the same had been done by the Treasury, or by the Comptroller-General or other officer of the Exchequer, or by the Warden of the Standards, or at the office of the Exchequer.

It shall be the duty of the Board of Trade to conduct all such comparisons, verifications, and other operations with reference to standards of measure and weight, in aid of scientific researches or otherwise, as the Board of Trade from time to time thinks expedient, and to make from time to time a report to Parliament on their proceedings and business under the Weights and Measures Act, 1878.

Powers and duties of Board of Trade as to standards of weights and measures, &c. 41 & 42 Vict. c. 49, s. 33.

> This section is for the most part a re-enactment of the provisions of the "Standard of Weights, Measures, and Coinage Act, 1866," which is now repealed (29 & 30 Vict. c. 82, ss. 1, 11, and 12). The enactment, however, for a separate office of Warden of the Standards and separate Standards Department is omitted, the existing arrangement whereby one of the secretaries of the Board of Trade is made Warden of the Standards without additional salary, and

the Standard Weights and Measures Department is made strictly a department of the Board of Trade, having been virtually sanctioned by Parliament.

Under the former Act, the Warden of the Standards made an annual report to the Board of Trade, which was laid before both Houses of Parliament. Now the report appears as the report of a department of the Board of Trade, and is not required to be annual.

<small>Custody of imperial and Board of Trade standards.
41 & 42 Vict. c. 49, s. 34.</small>

The imperial standards of measure and weight, the Board of Trade standards of measure and weight, and all balances, apparatus, books, documents, and things used in connection therewith or relating thereto, deposited on the eighth day of August 1878, in the Standards Department, or in any other office of the Board of Trade, shall remain and be in the custody of the Board of Trade.

<small>Custody and periodical verification of parliamentary copies of imperial standards.
41 & 42 Vict. c. 49, s. 35.</small>

The parliamentary copies of the imperial standards of measure and weight mentioned in Part Two of the First Schedule to the Weights and Measures Act, 1878, shall continue to be deposited as therein mentioned.

The copies of the imperial standards of measure and weight made in pursuance of the Weights and Measures Act, 1878, when approved by Her Majesty in Council, shall be deposited at some office of the Board of Trade, and be in the custody of the Board of Trade.

The Board of Trade shall cause the parliamentary copies of the imperial standards of measure and weight, except the copy immured in the new palace at Westminster, to be compared once in every ten years with each other, and once in every twenty years with the imperial standards of measure and weight.

<small>Periodical verification of Board of Trade standards.
41 & 42 Vict. c. 49, s. 36.</small>

Once at least in every five years the Board of Trade shall cause the Board of Trade standards for the time being to be compared with the parliamentary copies of the imperial standards of measure and weight made and approved in pursuance of the

Weights and Measures Act, 1878, and with each other, and to be adjusted or renewed, if requisite.

The Board of Trade shall cause to be compared with the Board of Trade standards and verified, at such place as the Board of Trade in each case direct, all copies of any of those standards which are submitted for the purpose by any local authority, and have been used or are intended to be used as local standards, and if they find the same fit for the purpose of being used by inspectors of weights and measures under the Weights and Measures Act, 1878, as standards for the verification and inspection of weights and measures, shall cause them to be stamped as verified or re-verified in such manner as to show the date of such verification or re-verification, and every such verification shall be evidenced by an indenture, and every such re-verification shall be evidenced by an indorsement upon the original indenture of verification, or by a new indenture of verification. *Verification by Board of Trade of local standards. 41 & 42 Vict. c. 49, s. 37.*

Any such indenture or indorsement, if purporting to be signed (either before or after the eighth day of August 1878) by an officer of the Board of Trade, shall be evidence of the verification or re-verification of the weights and measures therein referred to.

Any such indenture or indorsement shall not be liable to stamp duty, nor shall any fee be payable on the verification or re-verification of any local standard.

An account shall be kept by the Board of Trade of all local standards verified or re-verified.

The powers given to the Board of Trade to restore, or replace the imperial standards, the parliamentary copies of the imperial standards, and the Board of Trade, or secondary standards, have been considered in a previous chapter. See Standards of Measure and Weight, *ante.*

In the sections above, provision is made for the custody and periodical verification of all standards and copies. The imperial

C

standards are to remain in the custody of the Board of Trade, where also will be kept the parliamentary copies of the imperial standards which have been authorised by the Weights and Measures Act, 1878, to be made. These copies have been specially provided for the purpose of verifying with them the Board of Trade standards, and thus saving the frequent use of the imperial standards.

The parliamentary copies of the imperial standards in the hands of the authorities at the Mint, Royal Observatory, and Royal Society will remain in their custody, and the copies immured in the new palace at Westminster will remain in their position. The three former copies are to be compared once in every ten years with each other, and once in every twenty years with the imperial standards. Local standards of weight have to be verified every five years, and local standards of measure have to be verified every ten years, and any local standard which has become defective must be verified by the Board of Trade. See Local Administration, *post*. The Board of Trade can now fix the place of verification, and the stamp must show the date of the verification. The former Acts seemed to contemplate a comparison of the local standards with the imperial standards, but in practice they were always compared with the Board of Trade standards, and this practice is now distinctly authorised by the sections given above. As already noted the local standards will now be of the same shape as the Board of Trade standards.

The provision that the indenture, and any indorsement thereon, shall be evidence of the verification or re-verification of the weights and measures therein referred to, will clear up some doubts that have arisen, and render the proof of the legality of the standards a simple matter.

From the earliest known period of English history the standard weights and measures were kept at the Exchequer; and the duties relating to these standards were up to a comparatively recent period imposed on the Chamberlains of the Exchequer. Upon the abolition of the office of Chamberlain, in 1826, the duties connected with these standards were imposed upon the Auditor of the Exchequer. These duties were transferred to the Comptroller-General of the Exchequer in 1834, the actual performance of the duties being entrusted to a subordinate department of the Exchequer. By the Standards of Weights, Measures, and Coinage Act, 1866 (29 & 30 Vict. c. 82), the custody of the standards and of all balances and apparatus, as well as books, documents, and things used in connection with them, was, upon the Act taking effect (6th August), legally transferred to the Board of Trade, together with all powers and duties relating to those standards then by law vested in or imposed upon the Treasury, or the Comptroller-General of the Exchequer. The actual transfer of the custody of the standards, &c., and the determination of his powers and duties connected with them, were reported by the Comptroller-General of the Exchequer on the 31st of August, 1866, to the Treasury. See First Report of the Warden of the Standards, 1866-67.

Administration.

The Board of Trade, on payment of such fee, not exceeding five shillings, as they from time to time prescribe, shall cause all coin weights required by the Weights and Measures Act, 1878, to be verified, to be compared with the standard weights for weighing coin, and, if found to be just, stamped with a mark approved of by the Board, and notified in the London Gazette.

All fees under this section shall be paid into the Exchequer.

<small>Verification and stamping of coin weights. 41 & 42 Vict. c. 49, s. 39.</small>

<small>A similar provision for the verification and stamping of coin weights was made in the Coinage Act, 1870 (33 & 34 Vict. c. 10, s. 17). This is now repealed and the above enactment takes its place.

Required by the Weights and Measures Act, 1878, to be verified. See "Stamping and Verification of Weights and Measures," *post*.

There are other powers and duties of the Board of Trade, which will be found described elsewhere. Thus they have power from time to time to cause new denominations of standards to be made and duly verified, which shall be Board of Trade standards (41 & 42 Vict. c. 49, s. 8); they can alter the fees taken by the inspectors for comparing and stamping weights and measures (41 & 42 Vict. c. 49, s. 47); and they have to give their approval to the bye-laws which the local authority has power to make for regulating the verification and stamping of weights and measures in use within its jurisdiction, and the duties of the inspectors under the Weights and Measures Act, 1878 (41 & 42 Vict. c. 49, s. 53).</small>

(2.) *Local.—The Local Authority.*

The local authority of every county and borough from time to time shall provide such local standards of measure and weight as they deem requisite for the purpose of the comparison, by way of verification or inspection, in accordance with the Weights and Measures Act, 1878, of all weights and measures in use in their county or borough, and shall fix the places at which such standards are to be deposited.

The said local authority shall also provide from time to time proper means for verifying weights and measures by comparison with the local standards of

<small>Provision of local standards by local authority. 41 & 42 Vict. c. 49, s. 40.</small>

such authority, and for stamping the weights and measures so verified.

<small>Expenses of local authority.
41 & 42 Vict. c. 49, s. 51.</small>
The expense of providing and re-verifying local standards, the salaries of the inspectors, and all other expenses incurred by the local authority under the Weights and Measures Act, 1878, shall be paid out of the local rate.

The treasurer of the county in which a borough in England having a separate court of quarter sessions is situate, shall exclude from the account kept by him of all sums expended out of the county rate to which the borough is liable to contribute, all sums expended in pursuance of the said Act.

<small>Local authorities and local rate.
41 & 42 Vict. c. 49, s. 50.</small>
For the purposes of the Weights and Measures Act, 1878, "the local authority" and "the local rate" shall mean in each of the different areas mentioned in the first column of the Fourth Schedule to that Act the authority and the rate or fund mentioned in that schedule in connection with that area:

Provided that in England the council of a borough which has not a separate court of quarter sessions shall not, unless they so resolve, be the local authority for the purposes of the said Act, and if they so resolve and provide local standards and appoint inspectors after the commencement of the said Act, they shall forthwith give notice of such resolution and appointment, under the corporate seal of the borough, to the clerk of the peace of the county in which the borough is situate, and after the expiration of one month from the day on which that notice of the said appointment is given, the powers of inspectors of weights and measures appointed by the justices of the county shall, as to such borough and the weights and measures of persons residing therein, cease; but, until such notice is given, the borough shall be deemed to form part of the said county, in like manner as if the same were not a borough.

Where, at the commencement of the said Act, legal

local standards are provided, and inspectors are appointed by the council of a borough not having a separate court of quarter sessions, that council shall continue to be the local authority until they otherwise resolve.

Any two or more local authorities may combine, as regards either the whole or any part of the areas within their jurisdiction, for all or any of the purposes of the Weights and Measures Act, 1878, upon such terms and in such manner as may be from time to time mutually agreed upon. Power of local authorities to combine. 41 & 42 Vict. c. 49, s. 52.

An inspector appointed in pursuance of an agreement for such combination shall, subject to the terms of his appointment, have the same authority, jurisdiction, and duties, as if he had been appointed by each of the authorities who are parties to such agreement.

Shall provide.—The local authority is bound to provide such local standards as they may deem requisite, and proper means for the verification of weights and measures within its jurisdiction.

Local Standards.—These will be copies of the Board of Trade standards, and not as was implied by the former Acts copies of the imperial standards. See " Secondary or Board of Trade standards," *ante.*

By way of verification or inspection.—In this section it is made clear that the local standards are to be provided for the purpose of inspection as well as of verification.

Proper means.—This seems to give the local authority fuller powers than they had under the former Acts, which only provided for the procuring of stamps for the inspectors. Under the above phrase may be included a suitable office for the inspectors in crowded districts, where the extent of work to be done renders such accommodation advisable if not absolutely necessary.

All other expenses.—This also is a more general term than was used in the former statutes. It will remove any doubts that may have arisen with regard to many payments not hitherto distinctly authorised to be made, as it enables the local authority to pay any expense lawfully incurred under the new Weights and Measures Act.

The local authority—The local rate.—In the fourth schedule to the Weights and Measures Act, 1878, which will be found in the *Appendix*, the local authority of a county is declared to be the "Justices in general or quarter sessions assembled," and the local rate is the county rate. In the county of the city of London the local authority is declared in the same schedule to be the "Court of the Lord

Mayor and Aldermen of the city," and the local rate is the consolidated rate; and the local authority of a borough is the "mayor, aldermen, and burgesses acting by the council," the local rate being the borough fund and borough rate.

In the notes attached to the said schedule, the expression "county" is held not to include a county of a city or a county of a town, but to include every riding, division, or parts of a county having a separate court of quarter sessions. The Soke of Peterborough is to be deemed to be a county, but every other liberty of a county not forming part of the City of London, is to be deemed to form part of a county in which the same is situate, or which it adjoins; and if it adjoins more than one county, then of the county with which it has the longest common boundary. The expression "borough" means any place for the time being subject to the Municipal Corporation Act, 1835, and any Act amending the same, which has a separate commission of the peace.

By the Weights and Measures Act Amendment Act of 1859 (22 & 23 Vict. c. 56), authority was given to the town councils of all municipal boroughs incorporated under the Municipal Corporations Act, " to which a separate court of quarter sessions had been granted," to use and exercise within their respective boroughs all the powers concerning weights and measures, and the appointment of inspectors, &c., hitherto held and used by the court of quarter sessions of the county. Afterwards, in 1861, by the Municipal Corporations Act Amendment Act (24 & 25 Vict. c. 75), the above provisions were extended to all boroughs having a separate commission of the peace. By the present section the council of a borough which has not a separate court of quarter sessions has the option of being the "local authority." If they have provided standards and appointed inspectors up to the first day of January, 1879, they will continue to be the local authority until they otherwise resolve; if, on the other hand, they have not up to that time carried out the provisions of the former Weights and Measures Acts, they will not become the local authority under the Act of 1878 until they resolve to do so, and having provided standards and appointed inspectors, give the necessary notice to the clerk of the peace of the county.

The effect of the definitions given above will be to place the appointment of inspectors of weights and measures in all cases in the hands either of the magistrates of the counties or the council of the boroughs. Under the old system, in counties of cities and counties of towns where a court of quarter sessions had been granted, the Recorder, who had the powers of quarter sessions, appointed the inspectors (*R.* v. *Recorder of Hull*, 8 A. & E. 638; 2 J. P. 550); but he did not appoint them in places where they had formerly been appointed out of sessions (*Duly* v. *Sharood*, 3 Jur. N.S. 63; 20 J. P. 405 and 820).

Any two or more local authorities may combine.—This power to combine is a new power, except so far as it has existed to some extent in Scotland under the 17th section of 5 & 6 Will. IV. c. 63. It is a most useful provision, and will remove much inconvenience that has been felt in some localities in carrying out the provisions with regard to the law of weights and measures.

A local standard of weight shall not be deemed legal, nor be used for the purposes of the Weights and Measures Act, 1878, unless it has been verified or re-verified within five years before the time at which it is used. Periodical verification of local standards. 41 & 42 Vict. c. 49, s. 41.

A local standard of measure shall not be deemed legal, nor be used for the purposes of the said Act, unless it has been verified or re-verified within ten years before the time at which it is used.

A local standard of weight or measure which has become defective in consequence of any wear or accident, or has been mended, shall not be legal nor be used for the purpose of the said Act until it has been re-verified by the Board of Trade.

A local standard may, save as aforesaid, be re-verified, for the purpose of this section, by such local comparison thereof as is hereinafter mentioned, if on that local comparison it is found correct, but otherwise shall be, and in any case may be, re-verified by the Board of Trade.

A local comparison of a local standard shall be made by an inspector of weights and measures for the county or borough in which such standard is used, comparing the same, in the presence of a justice of the peace, with some other local standard which has been verified or re-verified by the Board of Trade, in the case of a weight within the previous five years, and in the case of a measure within the previous ten years.

Upon a local comparison where the local standard is found correct, the justice shall sign an indorsement upon the indenture of verification of that standard, stating such local comparison and verification, and the error, if any, found thereon, and the indorsement so signed shall be transmitted to the Board of Trade to be recorded in the account of the verification of local standards. The indorsement when so recorded shall be evidence of the local comparison and verification, and a statement of the record thereof, if purporting to be signed by an officer of the Board of Trade, shall be evidence of the same having been so recorded.

It shall be lawful for Her Majesty from time to time, by Order in Council, to define the amount of error to be tolerated in local standards when verified or re-verified by the Board of Trade, or when re-verified by such a local comparison as is authorised by this section.

<small>Production of local standards. 41 & 42 Vict. c. 49, s. 42.</small>

The local standards shall be produced by the person having the custody thereof, upon reasonable notice, at such reasonable time and place within the county, borough, or place for which the same have been provided, as any person by writing under his hand requires, upon payment by the person requiring such production of the reasonable charges of producing the same.

<small>These provisions for the periodical verification of local standards are for the most part re-enactments of similar provisions in the Acts of 1835 and 1859. There are, however, some important modifications. By section 5 of the 5 & 6 Will. IV. c. 63, it was enacted that all copies of the imperial standards "which may have become defective, or have been mended in consequence of any wear or accident, shall forthwith be sent to the Exchequer for the purpose of being again compared and verified." There was some doubt whether the word "forthwith" in this section meant forthwith after the passing of the Act, or forthwith after the copies have become defective. The present section adopts the more sensible construction, and provides that the defective standard shall not be legal, nor be used as a standard until it has been re-verified by the Board of Trade, without waiting until the expiration of the five or ten years when re-verification is required.

The cases in which a local verification of local standards is allowed have been largely extended by omitting the limitation that such re-verification was to be only in a county where there was more than one district for inspection, which was contained in section 1 of 22 & 23 Vict. c. 56. Such re-verification is now regulated by enacting that notice of it shall be sent to the Board of Trade, to be recorded by them, and that an indorsement of the verification shall be made on the indenture, and such indorsement is made evidence.</small>

The local comparison, however, is only to be effectual if the standard is found correct, and if it is not found correct, so that adjustment is required, re-verification by the Board of Trade will be necessary. The power given under 29 & 30 Vict. c. 82, s. 5, now repealed, to the Queen in Council to define the amount of error to be tolerated in comparing local standards with Board of Trade standards is in the present section extended to the case of a local comparison. The Order in Council, dated February 4th, 1879, defining the amount of error to be tolerated in local standards when verified or re-verified by the Board of Trade will be found in the *Appendix*. For the 'Gazette' notices in connection with coin weights, see page 86, *post*.

Local Standards shall be Produced.—This section is a re-enactment of a provision to a similar effect in the Act of 1824 (5 Geo. IV. c. 84, s. 12). It is the duty of the inspector to produce the standards if the requirements of the above section are properly met, and it would be a breach of duty rendering an inspector liable to a penalty under section 49 of the Weights and Measures Act, 1878, if he were to refuse or neglect to do so.

Any local authority from time to time, with the approval of the Board of Trade, may make, and when made, revoke, alter, and add to, bye-laws for regulating the comparison with the local standards of such authority, and the verification and stamping of weights and measures in use in their county or borough, and for regulating the local comparison of the local standards of such authority, and generally for regulating the duties under the Weights and Measures Act, 1878, of the inspectors appointed by the local authority or of any of those inspectors. Such bye-laws may impose fines not exceeding twenty shillings for the breach of any bye-law, to be recovered on summary conviction. The Board of Trade before approving any such bye-laws shall cause them to be published in such manner as they think sufficient for giving notice thereof to all persons interested.

<small>Power to local authority to make bye-laws as to local verification, &c. 41 & 42 Vict. c. 49, s. 53.</small>

This section is new, and was inserted in the Bill when passing through the House of Commons. The object of the section is to give additional powers to the local authority, and throw upon them the main responsibility of carrying the law into effect. The approval of the Board of Trade will be required of all bye-laws made by the local authority, and of all subsequent revocations and alterations. It is probable that some model bye-laws will be framed by the Board of Trade for adoption by the local authorities.

Appointment of inspectors in towns and other places. 41 & 42 Vict. c. 49, s. 54.	Where a town or other place has been or may hereafter be authorised under any Act, whether local or otherwise, to appoint inspectors or examiners of weights and measures, or where any other place has been or may hereafter be, by charter, Act of Parliament or otherwise, possessed of legal jurisdiction, and such town or place is for the time being provided with legal local standards, the magistrates of such town or place, or other persons authorised as aforesaid, may appoint inspectors of weights and measures within the limits of their jurisdiction, and suspend and dismiss such inspectors, and such inspectors shall within such limits exclusively have the same power and discharge the same duties as inspectors of weights and measures appointed under the Weights and Measures Act, 1878, by the local authority for the county, and shall pay over and account for the fees received by them under the said Act, to such persons as may be duly authorised by the magistrates or other persons appointing them.
Power of vestry, &c., in Metropolis to put an end to appointment of inspectors of weights and measures under Local Act. 41 & 42 Vict. c. 49, s. 55.	Where in any place in the Metropolis—that is to say, in the parishes and places in which the Metropolitan Board of Works have power to levy the consolidated rate—any vestry commissioners or other body have any duties or powers, under any Local Act, charter, or otherwise, in relation to the appointment of inspectors or examiners of weights and measures, such vestry commissioners or body may, at a meeting specially convened for the purpose, of which not less than fourteen days notice has been given, resolve that it is expedient that their said duties and powers should cease in such place.

The clerk or other like officer of such vestry, commissioners, or body, shall give notice of such resolution to the clerk of the peace for the county in which such place is situate, and the clerk of the peace shall lay such notice before the next practicable court of quarter sessions for the county, and after the receipt of such notice by the court of quarter sessions, the appointment, and all powers of appointment, of any inspector or examiner appointed under such Local Act, charter, or otherwise, shall cease in the said place, without prejudice to any proceedings then pending for penalties or otherwise.

There have been a considerable number of private, and local and personal, Acts passed from time to time, conferring special powers of inspection of weights and measures in various localities. The above section, taken from the Act of 1835 (5 & 6 Will. IV. c. 63. s. 25), reserves the power to appoint inspectors, &c., if the town or place is provided with legal local standards. It does not, however, create the "magistrates or other persons" a "local authority under the Act," an omission which is much to be regretted. These authorities will have power to provide standards and appoint inspectors, such inspectors having the same duties as those appointed by the local authorities: but for providing funds to pay for the same, for the verification of their standards, and for the making of regulations and bye-laws, they will have to depend upon the local Act or charter under which they act, and the powers therein given to them. Some difficulty may also arise with reference to an inspector appointed by the local authority for a county knowingly stamping a weight or measure of any person residing in the district of an inspector appointed under this section, as the penalty only applies to the intrusion of a county inspector into the district of an inspector "legally appointed by another local authority" (41 & 42 Vict. c. 49, s. 44), and thus an unseemly conflict between the two inspectors might arise, which it is the special object of the said prohibition to prevent. Such an intrusion, however, would, under the circumstances, probably be considered such "misconduct" upon the part of the county inspector as to bring him within the general provisions of section 49 of the Weights and Measures Act, 1878. See 41 & 42 Vict. c. 49, s. 49; and Inspectors and their Duties, *post*.

Examiners of Weights and Measures.—These officers were originally appointed under the provisions of the old Acts of 1795, 1797, and 1815, which, although only expressly repealed by the Act of 1878, have been practically useless and obsolete since the passing of the Act of 1835. These examiners, however, can still exist as separate officers in the parishes who have obtained their appointment under the old Acts; but as they cannot properly perform their duties without having local standards, and as it appears from the Standard Weights and Measures Department that no such standards have been furnished to examiners for many years, it is doubtful whether any such officers exist. If there are any in existence their rights are protected by section 86 of the Act of 1878 (41 & 42 Vict. c. 49, s. 86).

Suspend and dismiss.—This power was not given in the section of the Act of 1835, from which the present section was taken. It is here inserted to supply an obvious omission.

Vestry or other Body in the Metropolis.—There are at least four metropolitan parishes which have powers of appointment of inspectors of weights and measures under Local Acts or Charters: Paddington, Marylebone, St. Pancras, and St. Mary's, Islington. These powers, which differ in the various cases, have been used with considerable effect. In the Metropolis Local Management Act of 1862 (25 & 26 Vict. c. 102, s. 101), provision was made for these local authorities to determine their duties and powers with regard to weights and measures whenever they resolved to do so, and this provision is re-enacted in the section of the present Act given above.

The important duties which all local authorities have to perform in the appointment of inspectors of weights and measures will be found described in another chapter. See Inspectors and their Duties, *post*.

CHAPTER V.

THE USE OF WEIGHTS AND MEASURES.

The same weights and measures shall be used throughout the United Kingdom.

<small>Uniformity of weights and measures.
41 & 42 Vict., c. 49, s. 3.</small>

The earliest statutory regulations for the uniformity of weights and measures are contained in Magna Charta. One of the demands made by the Barons was that "there should be but one weight and measure throughout the kingdom;" and although, according to Blackstone, the regulation of weights and measures is the prerogative of the Sovereign, yet it has been continually subject to the control of Parliament. The provisions in Magna Charta relating to weights and measures are as follows: "There shall be but one measure of wine throughout the realm, and one measure of ale, and one measure of corn, that is to say, the quarter of London; and one breadth of dyed cloth, russets and haberjects, that is to say, two yards within the lists. And it shall be of weights as of measures."

Then followed in 1266 a statute for regulating measures, entitled "Assisa Panis et Cervisie," the "Assize of Bread and Ale," which signified the adjustment of the weights and measures of bread and ale, at the *sitting* of a competent tribunal. In 1709 the provisions of this older statute, with regard to the sale of bread, were repealed and fresh regulations of the assize of bread were made. Further legislation upon this subject followed until the assize of bread was finally abolished in 1822, in the metropolitan district, by the local Act, 3 Geo. IV. c. 106, and generally throughout the country in 1836, by the public Act, 6 & 7 Will. IV. c. 37. See The Sale of Bread and Coals, *post.*

In 1391, an Act was passed for the further regulation of measures used in buying corn, which was confirmed by another Act passed in 1413. These Acts recited the provisions in Magna Charta, that there was to be but one measure of corn throughout the realm, viz., the quarter of London, which quarter was to contain eight bushels stricken measure, and not nine bushels heaped measure. Any one buying corn otherwise than by the quarter of eight bushels stricken measure, was to be liable on conviction to a year's imprisonment, payment of a hundred pence to the king, and as much more to the party aggrieved.

In order to provide against the unlawful circulation of light or clipped coin, in addition to the ordinary weights, an apparatus called a tumbrel was, by a statute passed in 1292, to be employed for testing coin, to the value of five shillings; and both the tumbrel and weights were to be marked with the king's mark by an officer called the Warden of Exchange.

The aulnage of cloths, or the measurement and assize by the king's aulnager of manufactured cloths was ordained in 1328, the aulnager's fee and the aulnage duty being regulated in 1353. The aulnage of cloths was further regulated by a long series of enactments extending from the above date down to 1708. In 1719 an Act was passed regulating the dimensions not only of Scotch plaids, and serges, but of stockings. All stockings made in Scotland were to be made of three threads, of one sort of wool or worsted, and of three sizes; namely, for men called long stockings, for men called short stockings, and for women or boys. This Act was amended and extended in 1723.

The auncel weight, or hand-sale weight, was an ancient manner of weighing by the hanging of balances at each end of a staff, which, when lifted up with the hand in the middle, showed the equality, or difference, of the things weighed. This weight, being subject to much deceit, was abolished in 1350, and it was ordained that "every person should buy and sell by the even balance, and the wools and other merchandise be evenly weighed by light weight, so that the sack of wool should weigh no more than twenty-six stone, and every stone should weigh fourteen pounds, and that the beam of the balance should not bear more to one part than the other, and that the weight should be according to the standard of the Exchequer." If any buyer did the contrary he was to be "grievously punished," as well at the suit of the party as at the suit of the king. The auncel weight was the forerunner of the modern steelyards which are now so extensively used. The Act of 1350, above mentioned, was further confirmed and extended in 1353 and 1360.

In 1429 the Act 8 Hen. VI., c. 5, provided that every city, borough, and town should have a common balance with standard weights sealed according to the standard at the Exchequer; at which balance and weights all the inhabitants might freely weigh without paying anything. These common balances and weights were to be in the keeping of the mayor or constable, and be provided within two months under penalties; and in 1433 a common bushel, duly verified, was to be kept in each town, &c., upon the same plan.

The gauging of wine casks by the king's gaugers was directed in 1353, and further regulations with regard to the measure of wine were made by statutes passed between 1357 and 1706.

Regulations for the sale of cheese were passed in 1430, for the assize of barrels for fish in 1482, for the assize of herring casks in 1570, and for the sale by measure of honey in 1581.

Coal measure was first regulated in 1421, and provisions for the sale of coal, and the size of sacks, &c., are found in many statutes, down to the enactments prohibiting the sale of coal altogether by measure passed in 1835, and re-enacted in the schedule to the Weights and Measures Act, 1878. See The Sale of Bread and Coals, *post*.

The Winchester bushel, abolished in 1835, was defined in 1701 to be a "round bushel with a plain and even bottom, being eighteen and a half inches wide throughout, and eight inches deep" (13 Will. III., c. 5).

The "weight and packing of butter" is the subject of a series of regulations to be found in the statute book in Acts passed during a period extending from 1662 to 1830; and the sale of salt has also been regulated by Parliament, from 1670 down to 1798.

A measure chiefly used in maritime towns and known as "water measure," was described in 1701 as "round, and the diameter eighteen and a half inches within the hoop, and eight inches deep, and no more, and so on in proportion for a greater or lesser measure; and every measure, commonly called *water measure*, by which apples and pears are sold, shall be heaped as usually; and whoever shall sell or buy any apples or pears by and with any other measure, shall forfeit for each such offence ten shillings."

A "barrel of beer" was defined to contain "thirty-six gallons of beer taken by the gauge according to the Exchequer standard of the ale quart;" and a "barrel of ale" was to contain thirty-two such gallons (12 Car. 2. c. 23, s. 20). This Act was amended in 1688, and the barrel of either beer or ale was to consist of thirty-four gallons, according to the same standard. In 1699 a barrel of vinegar was to contain in like manner thirty-four gallons. But in 1803 the barrel of either beer or ale was to contain thirty-six gallons.

It was in 1824 by the 5 Geo. IV., c. 74, that all the provisions of former Acts were repealed relating to the ascertaining or establishing any standards of weights and measures, or to the establishing or recognising certain differences between weights and measures of the same denomination; and all weights and measures were to be determined and ascertained by the new imperial standards established under that Act. This Act is of course repealed by the Weights and Measures Act, 1878 (41 & 42 Vict. c. 49).

For a full description of the statutory regulations relating to weights and measures, see the Seventh Annual Report of the Warden of the Standards, 1872-73.

Every contract, bargain, sale, or dealing, made or had in the United Kingdom for any work, goods, wares, or merchandise, or other thing which has been or is to be done, sold, delivered, carried, or agreed for by weight or measure, shall be deemed to be made and had according to one of the imperial weights or measures ascertained by the Weights and Measures Act, 1878, or to some multiple or part thereof, and if not so made or had shall be void; and all tolls and duties charged or collected according to weight or measure shall be charged and collected according to one of the imperial

Trade contracts, &c., to be in terms of imperial weights and measures. 41 & 42 Vict. c. 49, s. 19.

weights or measures ascertained by the said Act, or to some multiple or part thereof.

Such contract, bargain, sale, dealing, and collection of tolls and duties as is in this section mentioned is in the said Act referred to under the term "trade."

No local or customary measures, nor the use of the heaped measure, shall be lawful.

Any person who sells by any denomination of weight or measure other than one of the imperial weights or measures, or some multiple or part thereof, shall be liable to a fine not exceeding forty shillings for every such sale.

<small>Sale by avoirdupois weight, with exceptions. 41 & 42 Vict. c. 49, s. 20.</small>

All articles sold by weight shall be sold by avoirdupois weight; except that—

(1.) Gold and silver, and articles made thereof, including gold and silver thread, lace, or fringe, also platinum, diamonds, and other precious metals or stones, may be sold by the ounce troy or by any decimal parts of such ounce; and all contracts, bargains, sales, and dealings in relation thereto shall be deemed to be made and had by such weight, and when so made or had shall be valid; and

(2.) Drugs, when sold by retail, may be sold by apothecaries weight.

Every person who acts in contravention of this section shall be liable to a fine not exceeding five pounds.

<small>Penalty on use or possession of unauthorised weight or measure. 41 & 42 Vict. c. 49, s. 24.</small>

Every person who uses or has in his possession for use for trade a weight or measure which is not of the denomination of some Board of Trade standard, shall be liable to a fine not exceeding five pounds, or in the case of a second offence ten pounds, and the weight or measure shall be liable to be forfeited.

<small>Exception for sale of article in vessel not</small>

Nothing in the Weights and Measures Act, 1878, shall prevent the sale, or subject a person to a fine under that Act for the sale, of an article in any vessel, where such vessel is not represented as containing any amount of imperial measure, nor subject a person to a fine under that Act for the possession of a vessel where it is shown that such vessel is not used nor intended for use as a measure.

represented as being of imperial measure. 41 & 42 Vict. c. 49, s. 22.

The above sections are perhaps the most important sections in the Weights and Measures Act, 1878. At all events they are the provisions to which a very large amount of attention has been directed, and they require some care in their consideration.

The first section has been described by a most competent authority* as "simply a re-enactment in altered and consolidated form of several clauses in the former Acts. Unless, therefore, the use of any denomination of weight or measure now in use for trade was illegal under the former law, the use of such denomination is not illegal under the new Act. The present custom of selling grain and other produce by weight or by measure, or by both weight and measure, is not affected by the Act." It may be as well, however, at once to point out that in one detail this description is not strictly accurate. For the 6th section of the Act of 1835 made it illegal to sell by any denomination of measure other than one of the imperial measures (5 & 6 Will. IV. c. 63, s. 6), and attached a penalty to such sale, no mention being made to the sale by weight, which, however, was, by the 11th section of the same Act, always to be by some multiple or aliquot part of the pound (5 & 6 Will. IV. c. 63, s. 11). It therefore sometimes became a question whether a sale was by measure or weight, and thus whether it was illegal or legal. The sale by "Hobbett" or "Hobbit," a measure much used in Wales, was declared to be illegal where the "Hobbit" was clearly a measure, and was not reducible to the legal standard (*Owens* v. *Denton*, 1 C. M. & R. 711; *Tyson* v. *Thomas*, 1 McClell & Y. 119). But where the "Hobbett" was defined to be 4 pecks of 42 lbs. each, a multiple of a pound, and it was in evidence that the delivery was always by weight, it was held to be a legal sale, and not contrary to the statute which applied to measures only (*Hughes* v. *Humphreys*, 23 L. J. Q. B. 356; 3 E. & B. 954; 1 Jur. N.S. 42; 2 W. R. 526; 23 L. T. 208; 18 J. P. 649). Now, however, in the present section the fine for selling by any measure other than imperial measure is extended to a sale by any weight other than imperial weight, and although this will not affect the decision in the case given above (*Hughes* v. *Humphreys*), where the weight was a multiple of an imperial weight, and so legal, yet there may be cases which have hitherto escaped being brought within the law owing to an accidental omission which has now been supplied.

The language of the first paragraph of this section has been made to expressly include the carriage of goods, and the collection of tolls and duties, matters which were before practically included

---

* Mr. Farrer of the Standards Department of the Board of Trade.

D

by the effect of sections 14, 21, and 28, of the Act of 1835. The result is to get the whole of these transactions included in the word "trade," and to obtain an inclusive definition of the term for use in other sections of the Act.

Local or Customary Measures.—These measures were abolished by section 6 of the Act of 1835 (5 & 6 Will. IV. c. 63, s. 6), and they are now declared to be unlawful. It is, therefore, somewhat disconcerting to know that "wheat at the present time is sold in this country in nearly fifty different ways. In one English county alone the term 'a stone' means upwards of thirty different weights. A bushel at Chester is seventy-five pounds; at Warrington it is seventy. There are twenty or thirty ways of measuring land; the pound weight varies in certain localities from sixteen to twenty-four ounces, and it is precisely the same with reference to every branch of trade."* It has also been stated that "there are not only different weights and measures for different things, but different weights and measures for different localities in the same county, and different weights and measures for different days in the same market town." How far these various weights and measures will be legalised or otherwise by the present Act will be considered in these notes; but it cannot be supposed that there are not many which, as "local or customary" measures, were abolished in 1835, and declared unlawful in 1878.

Heaped Measure.—This also was abolished in 1835, and is now declared unlawful (5 & 6 Will. IV. c. 63, s. 7). As, however, the penalty for selling by heaped measure has never been enforced, and is now expressly omitted, the illegality of such sale is comparatively unimportant.

Selling by other than Imperial Weight or Measure.—This is the section which has caused so much consternation among the trading communities, and it is a clause the correct meaning of which it is most important to ascertain. It has already been shewn that it is, for the most part, a re-enactment of a section in the Act of 1835, and that it has been declared upon authority that the only alteration in the law which was intended to be made is to attach a penalty to an improper sale by weight, which has hitherto been confined to an improper sale by measure. In dealing with the present provision, therefore, we are enabled to consider the manner in which the former enactment has been carried out in practice, and to obtain such assistance as we can find from any cases which may have already been decided upon this question.

And first, it is important to note that the offence consists in selling by *weight* or *measure* other than imperial weight or measure. It will therefore not include sales where no denomination of weight or measure at all is given. To measure or weigh anything is to determine its quantity according to some fixed standard. If therefore the quantity sold is merely a matter of conventional arrangement between the buyer and seller, the sale cannot be said to be

---

* The *City Press*, October 12, 1878.

by weight or measure at all. If there were any doubt upon this point it would be removed by the consideration of the exception provided for in the 22nd section, which expressly excludes from the operation of this prohibition the "sale of an article in any vessel, where such vessel is not represented as containing any amount of imperial measure."

We are thus enabled at once to deal with the important question of the sale of glasses of beer and wine, which sale, it has been suggested, will now become illegal. But wine-glasses and tumblers are not necessarily measures at all, and do not become so by being placed upon a refreshment bar; nor are they "represented as containing any amount of imperial measure." If a person asks for a glass of beer or a glass of wine, he asks for and obtains an indefinite quantity, which varies according to the size of the glasses; just as, if he asks for a cup of tea or coffee, he obtains an amount varying with the size of the cup. In neither case is the amount *measured* at all, nor is the sale a sale by measure. If half a pint of liquid is asked for, and it is served direct in a glass, then the glass becomes at once a measure, and must be stamped and contain half an imperial pint. The greatest care will be required in distinguishing between this "sale by convention" and a sale by measure, and of course the provisions of section 8 of the Licensing Act, 1872, which enacts that "all intoxicating liquor which is sold by retail, and not in cask or bottle, and is not sold in a quantity less than half a pint, must be sold in measures marked according to the imperial standard," must be duly observed (35 & 36 Vict., c. 94, s. 8).

Of a similar character to the sale of glasses of wine and beer is the sale of "pennyworths of milk" and "pennyworths of nuts," &c. Where the milk and the nuts are sold in mugs or other vessels, "not represented as containing any amount of imperial measure," the sale will not be by measure, and therefore will not be illegal. But where the delivery (as is often the case with milk) is made in a measure used for selling quantities of imperial measure also, then these measures must of course be correct.

The sale of beer, wine, spirits, and liqueurs, in bottle, will also for the most part come within the foregoing exceptions. As has been already explained, the bottle and half-bottle measures, which were declared to be "legal secondary standards of capacity" by Order in Council in 1871, are now no longer standards, as they were not included in the Schedule containing a list of the Board of Trade, or secondary standards. By the Order in Council of 1871, the bottle and half-bottle measures were declared to contain one-sixth and one-twelfth of a gallon respectively. Now, however, the bottles and half-bottles, if not used as measures, may be of any size. In most cases they will not be "vessels represented as containing any amount of imperial measure," but will be merely used as vessels to contain the liquids sold. In carrying out the provisions of the Weights and Measures laws a difficulty has often arisen in distinguishing between a mere vessel to contain liquids and a legal measure of capacity for liquids; and this difficulty is not likely to be removed by the present enactments. If persons are content to continue to buy wine merely by the dozen, there is no legal obligation on the seller to give either a dozen bottles, each containing one-sixth of a gallon, or a dozen bottles containing altogether two gallons. There is a further difficulty, arising from the fact that it would not be practicable to constitute every wine bottle a legal measure; nor if such a regulation were made applicable to bottles hereafter to be manufactured, could it be legally enforced, to say nothing of the immense inconvenience of doing away with the existing stock of bottles, and dealing with bottles of foreign manufacture containing wine and other liquors imported into this country. As a matter of fact the *exact* capacity of a wine bottle is a matter of chance in the manufacture, and it would be a practical impossibility to make all bottles of exactly the same size. The practice for the future will therefore be much the same as in the past. If a gallon, quart, or pint of beer, wine, or spirits is asked for it must be delivered in imperial measure, for there is only one gallon, quart, or pint known to the law; but if persons are content to order vaguely " bottles " and " dozens " of any liquid, they will continue to receive the indefinite quantity which has hitherto been delivered. There is, therefore, no chance of the sale of foreign liqueurs being illegal, or of old bottled port becoming a drug in the market. The above considerations will also apply to the sale of sauces, pickles, and scent, and all sales in vessels not " represented as containing any amount of imperial measure."

Again, a sale is permitted if the denomination of weight or measure is some " multiple or part of " an imperial weight or measure; and as the words " part of " were inserted in the place of the words " some aliquot part, such as the half, the quarter, the eighth, the sixteenth, or the thirty-second part thereof " during the passing of the Bill through Parliament, it is clear that we must take the words " multiple or part of " in the widest and most general signification. We must also carefully distinguish the prohibition in this case of selling from the stricter and more rigid prohibition of the use or possession of unauthorised weights and measures contained in the 24th section which follows. In that case, as will be shortly shewn, every weight or measure which is not of the denomination of some Board of Trade standard is illegal; but in selling by weight or measure *any* terms may be used so long as they are " some multiple or part of " an imperial weight or measure. If the terms of the imperial measures are used, then the law presumes that the imperial measures are meant. Thus a " bushel " taken by itself without reference to any particular agreement will mean an imperial bushel (*Hockin* v. *Cooke*, 4 T. R. 314); and in a lease the term " quarter of corn " means a legal quarter, containing sixty-four imperial gallons (*St. Cross* v. *Howard*, 6 T. R. 338).

But the most important decision bearing upon this point was that in the case of *Jones* v. *Giles* (10 Exch. 119 , where it was decided that a contract for the sale of a certain number of tons of iron by the ton " long weight," was not in contravention of the former statutes, and consequently that such a contract was valid. The ton "long weight" consisted of 2400 lbs., and was not the same as the ton defined in the Act of 1835 (which was the legal ton of 2240 lbs., as defined also in the Act of 1878, but it was the multiple of a pound, and as such was legally used in the contract. In giving judgment the judges held that all contracts for sale by weight must be by the standard pound, and if the words stone, hundredweight, and ton are used in a contract they must be taken conclusively to mean the weights mentioned in the Act; but if any other word or denomination of weight is used which, according to the ordinary rules of law applicable to the use of words, means a multiple of the standard pound, it is lawful and the contract is legal. This decision will help us to give a meaning to the present section. Under the former Act, upon which this decision was based, the terms stone, hundredweight, and ton were defined to be the same as in the present Act. But permission was given for any sale being made by any multiple or by some aliquot part of the pound weight. In the present sections the words are more general, for a sale may be by any "multiple or part" of any imperial weight or measure. So that, according to the above decision, if any of the terms of the imperial weights and measures are simply used they must be taken to mean the imperial weights and measures; but a sale may be by any term or denomination agreed to, so long as the denomination is some "multiple or part" of an imperial weight or measure.

Avoirdupois Weight.—It has already been noted (Chap. 1.) that the troy weight was the most ancient weight of this kingdom, having existed from the time of Edward the Confessor. The origin of the word has been said to have no reference to any town in France, but rather to the monkish name given to London of Troy Novant, founded on the legend of Brute. Troy weight, therefore, according to this etymology would be in fact London weight.* The avoirdupois weight arose by custom, but was also confirmed by statute. The two weights are thus described by an old writer: "Troy weight is by law; and thereby are weighed gold, silver, pearl, precious stones, silk, electuaries, bread, wheat, and all manner of grain or corn is measured by Troy weight. And this hath to the pound twelve ounces, or twenty shillings sterling weight and no more. It is called by some *Libra medica*; by others *Libra & uncia Trojana*. Averdupois weight is by custom (yet confirmed also by statute), and thereby are weighed all kind of grocery wares, physical drugs, butter, cheese, wax, pitch, tar, tallow, wools, hemp, flax, iron, steel, lead, and all other commodities not before named (as it seemeth), but especially every thing which beareth the name of garble, and whereof issueth a refuse, or waste. This is called *Libra civilis*. Averdupois in French is 'to have full weight,' *Habere pondus*. And this hath to the pound sixteen ounces or twenty-five shillings sterling weight." *Dalton*, c. 112. Another derivation has been given of the word avoirdupois, namely, from " Avoirs "

---

* McCulloch's 'Commercial Dictionary.'

(*Averia*), the ancient name for goods or chattels, and "poids," weight. The imperial standard pound was originally a brass weight of one pound troy weight made in 1758, and from this unit were derived the troy ounce, pennyweight, and grain, and from the grain were computed the pound, ounce, and dram avoirdupois. But in 1855 the imperial standard pound avoirdupois was declared to be the standard measure of weight from which were derived the other avoirdupois weights down to the grain, and from the grain was computed the pound troy. See Imperial Measures of Length, Weight, and Capacity, Chap. I., *ante*.

May be sold by the ounce Troy or by any decimal parts of such ounce.—The imperial standard pound avoirdupois contains seven thousand grains, and four hundred and eighty of these grains make up the ounce troy. As all sales by troy weight are now to be by the ounce troy, or by any decimal parts of such ounce, the troy pound and troy pennyweight are now practically abolished.

Other precious metals.—This is an extension of the sale of metals by troy weight, the former provision only including gold, silver, and platinum (5 & 6 Will. IV, c. 63, s. 10).

Apothecaries weight.—The standards of fluid ounces, fluid drams, and minims, have been added to the second Schedule of the Weights and Measures Act, 1878. As by that Act the weights and measures used by chemists and druggists for trade purposes are liable to inspection, Board of Trade standards of the apothecaries' measures of fluid ounces, fluid drams, and minims, have been constructed, verified, and declared legal by an Order in Council. Copies of these standards will be verified for the use of inspectors who will have to inspect the weights and measures used in druggists' shops.

Use or possession of unauthorised weights or measures.—Whatever may be said with regard to the preceding sections, the provisions of this clause are most definite and clear. The use or possession for use for trade of any material weight or measure which is not a copy of some Board of Trade standard is illegal. The section in the Act of 1835, upon which this clause is founded, was not so clear, as it prohibited "the use of any weight or measure other than those authorised by the Act." Not only was there great difficulty in proving the use of an unauthorised weight or measure, but it also became a question as to the true meaning of the words "unauthorised by this Act," and the practice of the inspectors throughout the country varied to a considerable extent. Now, however, the doubts are removed, and the practice can be uniform. The possession of a weight or measure, which is not a copy of one of the Board of Trade standards, will render a person liable to a penalty unless proof is given that the weight or measure is not for use for trade.

The list of Board of Trade standards will be found in the second Schedule of the Weights and Measures Act, 1878, given in the *Appendix*, and the definition of "trade" will be found in section 19 of the same Act, which has just been considered. See *ante*. Where any weight or measure is found in the possession of any person carrying on trade (as defined above), or upon the premises of any person which are used for trade (also as defined above), the weight or measure is to be deemed to be in the possession of that person for use for trade until the contrary is proved. See Legal Proceedings, *post*. A higher penalty can be imposed for a second offence under this section, if committed within five years from the time of the former conviction, and the unauthorised weight or measure can be seized by an inspector, and is liable to forfeiture.

There are four exemptions from the operation of this section which are either expressed or implied by the Weights and Measures Act, 1878. Thus, weights and measures may be kept for scientific purposes, being not used for trade, and as has already been shown, metric weights and measures may be used for the purpose of science or of manufacture. See Metric Weights and Measures, *ante*. There is also an exemption in favour of all weights and measures which have been duly verified and stamped at the passing of the Weights and Measures Act, 1878, and which may still be used, although they could not have been verified and stamped in pursuance of that Act (41 & 42 Vict. c. 49, s. 86). Lastly, by the succeeding section, permission is given for the "possession of a vessel, where it is shown that such vessel is not used, nor intended for use, as a measure."

In considering this last exception, which appears for the first time in the Act of 1878, many of the observations already made with regard to the true construction of the 19th section will apply. As to whether a vessel is used or intended to be used as a measure will in all cases be a question of evidence, the burden of proof resting upon the defendant. Where a vessel is not represented as containing any amount of imperial measure, or where, in fact, it does not contain any definite quantity judged by any fixed standard, it can hardly be considered a measure at all, but will be merely a vessel to contain the liquid which is to pass from the seller to the purchaser.

It should be noted that this section applies only to material weights and measures, and not to a weight or measure used merely as a term in a contract. Material weights above fifty-six pounds are rarely used, and if any demand arises for them the Board of Trade will be able to at once make a new standard for the purpose. The Board of Trade have already exercised their powers in this direction by making the *cental*, or 100 lbs. weight, a legal secondary standard of weight, in accordance with the provisions of section 8 of the Act of 1878. See Order in Council dated February 4th, 1879, in the *Appendix*.

Any person who prints, and any clerk of a market or other person who makes, any return, price list, price current, or any journal or other paper containing price list or price current, in which the denomination of weights and measures quoted or referred to denotes or implies a greater or less weight or measure than is denoted or implied by the same

Penalty on price lists, &c., denoting greater or less weight or measure than the same denomination of imperial

weight or measure.
41 & 42 Vict. c. 49, s. 23.

denomination of the imperial weights and measures under the Weights and Measures Act, 1878, shall be liable to a fine not exceeding ten shillings for every copy of every such return, price list, price current, journal, or other paper which he publishes.

This section, to which so much attention has been drawn by the unfavourable criticism of many of the most important newspapers in the country, is simply a re-enactment of a section in the Act of 1835 (5 & 6 Will. IV. c. 63, s. 31). There is, therefore, no change in the law which has been in force for the last forty-three years. As some misconception seems to have arisen with regard to the meaning of this section, it should be observed that the offence mentioned does not consist in quoting a weight or measure unknown to the law, but in quoting a weight or measure known to and determined by the law with a meaning different from that given to it by the law. Thus denominations of weights and measures which are not set forth in the Weights and Measures Act, if used or quoted, will not be affected by this section. But if such terms as bushel or quart, for example, are used, it will be illegal to refer to or quote them unless they contain imperial measure. See notes attached to the preceding sections, *ante*.

CHAPTER VI.

UNJUST WEIGHTS, MEASURES, AND WEIGHING MACHINES.

To sell by false weights and measures is an offence at the common law known by the name of Cheating. As such it is an indictable offence, and may be punished by fine and imprisonment. The law was thus laid down by Lord Mansfield in the case of *R.* v. *Wheatley*, 1 Bla. Rep. 273; 2 Burr. 1125, 1130: "Private unfair dealings which do not affect the public are not indictable offences *unless accompanied with false weights and measures*. All indictable cheats are where the public in general may be injured, as by using false weights, measures, or tokens." So in the case of *R.* v. *Young*, 3 T. R. 104: If a person sell by false weights, though only to one person, it is an indictable offence; but if, without false weights, he sells to many persons a less quantity than he pretends to do, it is not indictable. See also *R.* v. *Dunnage*, 2 Burr. 1130; and *R.* v. *Eagleton*, Dears, C.C., 376, 515.

But the law has for a long time provided a more summary procedure for the punishment of using unjust weights and measures than by indictment. Hitherto, however, the different sections in various Acts of Parliament for the purpose have been very varied and complicated, and an effort has been made in the following sections of the Weights and Measures Act, 1878, to simplify the existing enactments upon the subject.

Every person who uses or has in his possession for use for trade any weight, measure, scale, balance, steelyard, or weighing machine, which is false or unjust, shall be liable to a fine not exceeding five pounds, or in the case of a second offence ten pounds, and any contract, bargain, sale, or dealing, made by the same shall be void, and the weight, measure, scale, balance, or steelyard, shall be liable to be forfeited.

<small>Penalty on use or possession of unjust measures, weights, or weighing machines. 41 & 42 Vict. c. 49, s. 25.</small>

Where any fraud is wilfully committed in the using of any weight, measure, scale, balance, steelyard, or weighing machine, the person committing

<small>Penalty for fraud in use of weight,</small>

such fraud, and every person party to the fraud, shall be liable to a fine not exceeding five pounds, or in the case of a second offence ten pounds, and the weight, measure, scale, balance, or steelyard, shall be liable to be forfeited.

<small>measure, balance, &c. 41 & 42 Vict. c. 49, s. 26.</small>

A person shall not wilfully or knowingly make or sell, or cause to be made or sold, any false or unjust weight, measure, scale, balance, steelyard, or weighing machine.

<small>Penalty on sale of false weight, measure, balance, &c. 41 & 42 Vict. c. 49, s. 27.</small>

Every person who acts in contravention of this section shall be liable to a fine not exceeding ten pounds, or in the case of a second offence, fifty pounds.

These sections are probably the most important in the Weights and Measures Act, 1878. They are, at all events, those which affect the largest number of persons, and which will, it is to be feared, be the oftenest brought into requisition and referred to.

The former law upon the subject, which was extremely complicated, has been strengthened and amended in several important details by the foregoing sections, which, for the most part, follow the most stringent of the former enactments, namely, the one applying to persons selling in the streets. The provisions for preventing the use of unjust weights and measures when selling in shops and when selling in the streets have been amalgamated, and the 25th section, given above, renders a person liable for using or having in possession for use an unjust weight, measure, or scale *wherever it may be.*

Uses or has in his possession for use for trade.—This is an important phrase, introduced, for the first time, in the Act of 1878. The definition of "trade" has already been given in the 19th section of the Act (See The Use of Imperial Weights and Measures, *ante*), and it is wide enough to cover nearly all commercial dealings. Under the former Acts there was often some difficulty in determining whether the mere possession of an unjust weight or measure, without any proof of user, rendered the person in whose possession it was found liable to a penalty. It was of course a question of evidence for the justices to determine in each case. But these difficulties will now, for the most part, be removed, as it is distinctly enacted that the possession of a weight, measure, or scale, &c., is to be presumed to be for use for trade, until the contrary is proved. See Legal Proceedings, *post*. With the burden of proof upon the defendant, it will still be a question of evidence in each case, upon which the magistrates will have to decide. The character of the business done, and especially the position in which the weights, measures, or scales were found, will of course be important matters for consideration.

Weight, Measure, Scale, Balance, Steelyard, or Weighing-machine.—This phrase is a combination of the terms used in the former Acts. Thus in 1795 and 1797 we have the words, "balance or balances;" in 1835 we have "steelyards or other weighing-machines;" and in 1859 we find "beams, scales, or balances." In the present section the word "weighing-machine" is used in its strictly technical sense, and denotes the fixed machines used for the determination of very heavy weights at railway stations, wharves, &c.; whilst the words "scale, balance, and steelyard," will probably include all the ordinary weighing-machines used in shops and warehouses, the "balance" being strictly a machine constructed with the greatest care, adjusted with extreme nicety, and used for investigations which require more than ordinary accuracy.

False or Unjust.—The words in the former Acts were "light or otherwise unjust," which left it somewhat doubtful as to whether the use of a *heavy* weight was illegal. This doubt is now quite removed by the omission of the word "light" and the substitution of the word "false." These words also apply now to the use of scales and machines; the words of the former Acts in this case having been "incorrect or otherwise unjust." But in the 3rd section of the Act of 1859, with regard to selling in the streets (22 & 23 Vict. c. 56, s. 3), the inspector had power to inspect all beams, scales, and balances, and weights and measures in the possession of any person *selling, offering, or exposing for sale* any goods on any open ground, or in any public street, lane, thoroughfare, or other open space; and if upon such inspection or examination any such beams, scales, or balances, or weights or measures, were found to be *light or unjust*, . . . the same were liable to be seized and forfeited, and the person using or having them in his possession was liable to a penalty not exceeding five pounds." It will be seen that according to the wording of this section it was to apply to persons "selling, or offering or exposing for sale," their goods in the streets. So that in a case where a person was found selling in the street with a scale *against the seller*, and where fraud in the use of the scale was not alleged, it was held that there could be no conviction under this section, as it applied only to persons *selling* in the street, and if they sold only with a machine against themselves it could not be said to be false or unjust. The section just quoted was intended to protect the public against persons *selling*, and did not apply to the case of a seller using a balance unjust to himself, there being no provision for the unusual case of a person going out into the streets to buy and cheating the seller (*Booth* v. *Shadgett*, L. R. 8 Q. B. 352; 42 L. J. M. C. 98; 29 L. T. N.S. 30; 21 W. R. 845; 37 J. P. 743). The foregoing decision only applied to cases under the section of the Act of 1859, against persons selling in the streets, and had no reference whatever to the use of scales against the seller in shops, &c., although it has often been quoted as a general decision in favour of the use of scales against the seller. It is obvious that this was not so, and that in the case of persons who buy as well as sell, such as marine store dealers and others, the scales are unjust if they are against the seller, and the weights are unjust if they are too heavy. However the matter is thoroughly cleared up in the present section, which combines the provisions of the two former ones; and just as a weight, measure, or scale is "false or unjust," if found to be light, small, or against the purchaser where used for the purpose of selling, so it will be equally "false or unjust" if found to be heavy, large, or against the seller, where used for the purpose of buying.

In the case of a second offence ten pounds.—This provision is new. The second offence must be committed within five years from the time of the former conviction. For the method of proving the previous conviction and for the procedure generally, see Legal Proceedings, *post*.

Weight, measure, scale, balance or steelyard, shall be liable to be forfeited.—This renders every unjust weight, measure, or scale, liable to forfeiture, and consequently, liable to seizure by the inspector. See Inspectors and their Duties, *post*. The unjust weighing machines are excluded from this provision for the simple reason that they are, as a rule, fixtures, and practically could not be seized and forfeited. Under the former Acts there was no power to seize a scale, except where it was used in selling in the streets, so that this new enactment is a most important one. See *Thomas v. Stephenson*, 22 L. J., Q. B., 258; 2 E. & B., 108; 17 Jur. 597; 1 W. R., 325; 17 J. P., 537. The weights, measures and scales, &c., are only "liable to be forfeited" under this section, and they "may" be seized by the inspectors, so that there is a discretion in the justices in the one case, and in the inspectors in the other. The forfeiture is not a necessary consequence of a conviction, and so it should be expressly declared and contained therein.

It is, however, only in extreme cases that the power to seize a scale is likely to be exercised. There are some scales so old, and constructed in such a manner, that they bring their owners into trouble over and over again; and here it would be not only a protection to the community but a benefit to the defendant, to seize and forfeit a machine which is incapable of being made to work correctly. Again, there will occasionally be cases of gross fraud, where the machine is wilfully and designedly made for the purpose of cheating, and where this power can most usefully be put in force.

Fraud in use of weight, measure, scale, &c.—This is an extension of a provision which, in the former Act of 1859, only applied to persons selling in the streets (22 & 23 Vict. c. 56, s. 3.). It is now extended to persons selling in shops, and to fraud in the use of steelyards and weighing machines. This section also makes "every person party to the fraud" liable to the penalty, and not only the person using or possessing the weight, measure, or scale, and gives a higher limit to the penalty in the case of a second offence committed within five years from the time of the former conviction. This section will be useful to meet the cases, which, however, seldom occur, where the weight, measure, or scale being itself accurate is used in such a manner as to become even more fraudulent than if it were incorrect and unjust. In most cases the fraudulent use of a weight, measure, or scale, is in addition to the inaccuracy of the weight, measure, or scale which is used, and can therefore be met by a conviction under the preceding section. There are, however, some cases where it is not so, and it is to meet these that the present section has been framed. For example, the placing of a weight out of sight underneath say, the handle of a coal scoop, so as to give short measure, the scale being perfectly correct and just when the weight is removed, or throwing one of the chains of a scale over the beam and thus giving short weight to the customer, the scale being otherwise correct, would both be cases to be dealt with under this section.

Wilfully or knowingly make or sell, &c.—This is a re-enactment of a similar section in the Act of 1859 (22 & 23 Vict. c. 56, s. 2). It is, however, extended to the making or sale of steelyards and weighing machines, and a higher penalty is awarded in the case of a second offence committed within five years from the time of the former conviction. There is no power to seize or forfeit these unjust weights, measures, or scales, &c., until they are sold and come into the possession of a person for use for trade.

With regard to the number of summonses to be issued in cases where more than one unjust weight, measure, or scale is found in the possession of a person at the same time, a very important question arises from the alterations which have been made in the wording of the new sections upon this subject. In section 28 of the Act of 1835, "the person or persons in whose possession weights or measures, which were light or otherwise unjust were found, were, on conviction, to forfeit a sum not exceeding five pounds; and any person who had in his possession a steelyard or other weighing machine which was incorrect or otherwise unjust was to be liable to a like penalty" (5 & 6 Will. IV., c. 63, s. 28.). Although the practice has differed in different divisions, the most usual plan has hitherto been to follow the reading of the above section strictly and bearing in mind the provision of section 10 of the Summary Jurisdiction Act, 1848, that "every information shall be for one offence only" (11 & 12 Vict. c. 43, s. 10), the custom has been to issue one summons to include any number of weights, one summons to include any number of measures, and one summons for each weighing machine found to be unjust. In some divisions the weights and measures have been included in one summons, and in others more than one weighing machine have been taken together, but the most usual practice has been to combine the weights and combine the measures, but separate the machines. Under the sections of the present Act the practice is not so clearly laid down as might be desired, and the number of summonses to be issued in the case of a person found say, in the possession of two unjust scales, three unjust measures, and four unjust weights, is a question not easy to determine. Although the Act is declared officially to be a consolidating one, the practice here will most distinctly be altered, for at all events it is clear that in the case we have just supposed, there must either be one summons or nine summonses, the person having committed either one offence or nine offences. The words "any weight, measure, scale, &c.," would seem to imply that the possession of each weight, measure or scale was a distinct offence, and consequently would require a distinct information and summons; but when it is remembered that in the consideration of statutes the singular includes the plural, that penal statutes are to be construed strictly, and that in the analogous case of larceny the offence is not legally greater in proportion as the amount stolen is larger, it would seem to be the safer construction, as it will undoubtedly be the more reasonable practice, to treat the possession of any number of weights, measures, or scales as one offence, and to deal with it accordingly. The limit of the penalty which may be awarded, together with the new provisions, giving a higher limit in the case of a second offence, will leave a sufficient margin to enable a distinction to be made in the punishment of offenders in proportion to the extent of the offence which they may have committed.

The following are the most important cases which have been decided with reference to the use of unjust weights, measures, and scales, &c.

A police station where there were weighing machines and weights for the purpose of weighing coals which were allowed to the constables, was held not to be a "place where goods were exposed and kept for sale" under section 28 of 5 & 6 Will. IV. c. 63, and the weights so used were not within the 21st section of the same Act. There must be buying and selling—the coals were given to the constables. Nor were they "weighed for conveyance or carriage" (*Wray* v. *Reynolds*, 1 E. & E. 165; 22 J. P. 753).

It is doubtful whether this case would come within the provisions of the new Act, and whether such division of coals for delivery to the constables, as part of their pay, it is presumed, would be within the definition of the term "trade."

A machine, which from its construction was liable to variation from atmospheric and other causes, and required to be adjusted before it was used, was not incorrect upon examination within the meaning of the 28th section of 5 & 6 Will. IV. c. 63, if examined by the inspector before it had been adjusted. When adjusted, the machine was correct, and it was held to be going too far to say that because a machine requires adjustment before it is used it is an incorrect or otherwise unjust machine (*L. & N. W. Railway* v. *Richards*, 2 B. & S. 326; 8 Jur. N.S. 539; 5 L. T. N.S. 792; 26 J. P. 181).

But where a machine is out of repair, and does not require merely adjusting, but mending, the conviction was affirmed. A machine must be correct in itself, and not merely capable of being used so as to weigh correctly by making allowance for the difference; and even if it is shewn that the machine is used so as to do no actual injustice to the customer, it is incorrect within the meaning of the statute (*G. W. Railway* v. *Bailie*, 34 L. J. M. C. 31; 5 B. & S. 928; 11 L. T. N.S. 418; 11 Jur. N.S. 264; 13 W. R. 203; 29 J. P. 229).

A balance ball has been held under certain circumstances to be illegal. Where the ball was easily removable, being hung on simply, it was deemed to be an instrument of adjustment, and not an integral part of the machine, which without it *might* consequently be considered unjust. In this case, however, the balance ball formed no part of the scale, and it was easily detached. If used at all the balance ball should be fixed on to the machine so as not to be removable, and if so fixed it becomes the most convenient method of keeping a scale correct (*Carr* v. *Stringer*, L. R. 3 Q. B. 433; 37 L. J. M. C. 120; 18 L. T. N.S. 399; 9 B. & S. 238; 16 W. R. 859; 32 J. P. 307, 517).

A relieving officer who uses a weight to check the delivery of bread supplied to a Union for the purposes of out-door relief, is liable to be convicted if the weight so used is unjust (*Painter* v. *Seers*, 40 J. P. 549).

A coal merchant put some pieces of wood called a " Barrow " on a scale for the convenience of weighing. This barrow weighed 5 lbs., and the sacks weighing 2 lbs., a piece of iron weighing 7 lbs. was put on the other side of the scale to make it even. This was done in the presence and with the consent of the purchaser, and there was no intention to do wrong. Here it was held that no wrong was done, and that the scale was not incorrect (*Withall* v. *Francis*, 42 J. P. 84, 612).

The prosecution of these offences and the recovery of the fines will be carried out in accordance with the provisions of the Summary Jurisdiction Act, 11 & 12 Vict. c. 43. For a description of the principal provisions of that Act, together with suggestions having reference to points which are likely to arise in practice, see the chapter upon Legal Proceedings, *post*.

CHAPTER VII.

STAMPING AND VERIFICATION OF WEIGHTS AND MEASURES.

<small>Stamping of weights and measures with denomination.
41 & 42 Vict. c. 49, s. 28.</small>

Every weight, except where the small size of the weight renders it impracticable, shall have the denomination of such weight stamped on the top or side thereof in legible figures and letters.

Every measure of capacity shall have the denomination thereof stamped on the outside of such measure in legible figures and letters.

A weight or measure not in conformity with this section shall not be stamped with such stamp of verification under the Weights and Measures Act, 1878, as is hereinafter mentioned.

<small>Stamping of verification on measures and weights.
41 & 42 Vict. c. 49, s. 29.</small>

Every measure and weight whatsoever used for trade shall be verified and stamped by an inspector with a stamp of verification under the Weights and Measures Act, 1878.

Every person who uses or has in his possession for use for trade any measure or weight not stamped as required by this section, shall be liable to a fine not exceeding five pounds, or in the case of a second offence ten pounds, and shall be liable to forfeit the said measure or weight, and any contract, bargain, sale, or dealing made by such measure or weight shall be void.

<small>Lead or pewter weights.
41 & 42 Vict. c. 49, s. 30.</small>

A weight made of lead or pewter, or of any mixture thereof, shall not be stamped with a stamp of verification or used for trade, unless it be wholly and substantially cased with brass, copper, or iron, and legibly stamped or marked "cased":

Provided that nothing in this section shall prevent the insertion into a weight of such a plug of lead or pewter as is *bonâ fide* necessary for the purpose of adjusting it and of affixing thereon the stamp of verification.

A person guilty of any offence against or disobedience to the provisions of this section, shall be liable to a penalty not exceeding five pounds, or in case of a second offence ten pounds.

Where a measure for liquids is constructed with a small window or transparent part through which the contents, whether to the brim or to any other index thereof, may be seen without impediment, such measure may be verified and stamped by inspectors under the Weights and Measures Act, 1878, although such measure is made partly of metal and partly of glass or other transparent medium, and that whether such measure corresponds exactly to a Board of Trade standard, or whether it exceeds such standard, but has the capacity of such standard indicated by a level line drawn through the centre of the window or transparent part.

<small>Power to stamp measures made partly of metal and partly of glass. 41 & 42 Vict. c. 49, s. 46.</small>

In re-enacting the various provisions with regard to the stamping and verification of weights and measures in the former Acts, the opportunity has been taken in the foregoing sections to make several necessary modifications in the law, and to supply some important omissions.

In the Act of 1835 the stamp noting the denomination of a weight was only required upon weights made after the passing of that Act, and of the weight of one pound and upwards. The former exception is now omitted as practically unnecessary, and the stamp is required upon all 'weights "except where the small size of the weight renders it impracticable."

Shall not be Stamped.—This refusal to stamp by the inspector is the only result of disobeying the provisions of the previous clauses, there being no penalty attached to the same. As, however, this refusal will render the weights and measures practically useless, for they can only be used at the risk of incurring seizure and forfeiture for being unstamped, an additional penalty was not required. This result has been attained by extending the refusal to stamp, which under the former Acts was confined to troy weights only, to all weights and measures (16 & 17 Vict. c. 29, s. 5).

As is hereinafter mentioned.—This refers to the provision for stamping by an inspector contained in the section immediately succeeding, which is given above exactly in the place which it occupies in the Act itself.

Used for Trade.—This phrase refers us back again to the definition of "trade" given in the 19th section of the Act (See The Use of Imperial Weights and Measures, *ante*), and makes it clear that weights and measures used for other purposes, such as science or manufacture, need not be stamped in accordance with the requirements of this section.

Verified and Stamped.—The stamping is in all cases to follow the verification, when upon examination the weight or measure is found correct. "Stamping" is defined in the 70th section of the Weights and Measures Act, 1878, to include "casting, engraving, etching, branding or otherwise marking, in such manner as to be so far as practicable indelible" (41 & 42 Vict. c. 49, s. 70).

Has in his Possession for Use for Trade.—This is a very important addition to the provisions of the former Acts. Under section 21 of the Act of 1835, a person was only liable for *using* an unstamped weight or measure, and the difficulty of proving the user rendered the enactment almost inoperative (5 & 6 Will. IV. c. 63, s. 21). Now, however, the *possession* for use for trade of an unstamped weight or measure will render a person liable to a penalty, and the possession will be deemed to be a possession for use for trade until the contrary is shown. See Legal Proceedings, *post*.

In the case of a Second Offence, Ten Pounds.—This increase to the limit of the penalty in the case of a second conviction is also an addition to the former provisions. The conviction must be within five years from the time of the former conviction. For methods of proving previous convictions and procedure generally, see Legal Proceedings, *post*.

Shall be liable to Forfeit.—The liability of seizure and forfeiture of an unstamped weight or measure is again a provision which appears for the first time in the Act of 1878. It has been, however, inserted to supply an obvious omission in the section of the Act of 1835, dealing with the same subject (5 & 6 Will. IV. c. 63, s. 21).

The 21st section of the Act of 1835, which has already been so often referred to, contained two important exceptions which have been omitted in the present enactment.

In the first of these, no single weight above fifty-six pounds, that being the greatest of the imperial standard weights at that time deposited in the exchequer, was required to be inspected and stamped. Now, however, no material weight heavier than fifty-six pounds is legal, unless the Board of Trade, finding that such heavy weights are required for trade, legalise additional standards under the 8th section of the Act of 1878, by which these weights can be compared and stamped in the same manner as weights of a smaller denomination (41 & 42 Vict. c. 49, s. 8). The 100 lbs. weight or "cental" has already been legalised in this manner. See Order in Council, dated February 4th, 1879, in the *Appendix*.

The second exception was in favour of wooden or wicker measures used in the sale of lime or other articles of a like nature, and glass or earthenware jugs or drinking cups, which also were not required to be stamped, although persons using these earthenware jugs were bound to compare them with duly stamped measures it required by the buyer to do so. The difficulty of stamping these measures was primarily the reason of this provision; but as, owing to the extended definition of the word "stamping" in the 70th section of the present Act, this difficulty no longer exists, the exception is now omitted.

Earthenware vessels having been exempted by the 21st section of the Act of 1835, from being stamped, it became a question whether, if used as measures, they were liable to be seized and forfeited if found to be unjust under section 28 of the same Act. It was held that earthenware vessels, although they might not be stamped, yet if they were represented as containing a certain quantity according to imperial measure and were ordinarily used as measures, were liable to be seized and forfeited, if upon examination they were found to be unjust (*Washington* v. *Young*, 5 Exch. 403; 19 L. J. N.S. Ex. 348; 15 L. T. 234; 11 J. P. 591; confirmed by *R.* v. *Aulton*, 3 E. & E. 568; 30 L. J. M. C. 129; 7 Jur. N.S. 238; 3 L. T. N.S. 699; 9 W. R. 278; 25 J. P. 69).

Now all vessels which are represented to contain any quantity of imperial measure, and which are used as measures, must be stamped both with the stamp of denomination and that of verification.

It is not necessary where a weight has been once stamped, but from use the mark has become erased, that it should be restamped; and if there is reliable evidence that the weight has been stamped, there could not be a conviction for the possession of such a weight (*Starr* v. *Stringer*, L. R. 7 C. P., 383; as *Starr* v. *Trinder* in 26 L. T. N.S., 735).

But this decision is not likely to be of very much importance in ordinary cases, as the use and wear of a weight that would erase the stamp, would in most cases also make the weight incorrect, and so render necessary adjustment and restamping.

Lead or Pewter weights.—This section, although the wording is not quite the same, is practically the re-enactment of a similar provision in the Act of 1835 (5 & 6 Will. IV., c. 63, s. 13). The last clause imposing a penalty for the infringement of the provisions of the section has however been added, probably because the previous enactments have never been fully enforced. The law permits " the insertion into a weight of a plug of lead or pewter, which is *bonâ fide* necessary for the purpose of adjusting it and affixing thereon the stamp of verification." Under the former Acts considerable advantage was taken of this permission, which was interpreted in its widest sense, and a piece of lead was often placed upon the top of the weight and technically called the "cap," into which the adjusting plug of lead was inserted. This practice, if the present section is construed strictly, is however illegal; the only lead which is permitted being the plug which is inserted into the weight for the purpose of adjusting it and stamping it with the stamp of verification.

It has been suggested that as the plug can only be used for adjusting *and* stamping, it necessarily follows that every time that the plug is so used for adjusting the weight it must be restamped and generally, that no weight can be readjusted without being restamped. With regard to the first point, if the provisions of the section are strictly carried out, and the lead is confined to the plug inserted into the weight which is used for adjusting and stamping, any readjusting of the weight must almost necessarily interfere with the stamp and render a restamping absolutely indispensable: and in practice no one would venture to readjust the weight with the plug himself, nor do the more respectable scalemakers do so without sending the weight to the inspector for verification and restamping.

As to the general question, adjusting may mean anything from the simple washing and cleaning of dirty weights and measures to the most thorough and complete restoration. Few persons are likely to undertake the adjusting of their own weights and measures, but there is nothing to prevent their doing so if they please, and as often as may be found necessary. Nor is there any direct enactment providing that whenever this adjusting takes place, the weight or measure is to be taken to the inspector to be verified and restamped. It is obvious that in the case of such a simple adjusting as cleaning, such a provision would be ridiculous. If a weight or measure is stamped and correct, it is difficult to see under what section a person can be summoned for adjusting his own weights and measures and not having them restamped. He must not, of course, "wilfully increase or diminish a stamped weight" (see s. 32, *post*), and if in attempting to adjust his weights and measures himself he makes them more incorrect than they were before —as will often be the case—he will of course be liable for having unjust weights and measures in his possession. This difficulty is more a theoretical than a practical one. In practice the weights and measures are sent to a respectable scalemaker, who, having adjusted them, sends them as a matter of course to be verified and restamped. If a fraudulent scalemaker under the pretence of putting weights and measures in order, does the reverse, he will not only be liable to the penalties imposed by the 32nd section of the Act for "illegally increasing or diminishing a stamped weight," but he might be indicted for obtaining money by means of false pretences.

Measures partly made of Metal and partly of Glass. This is mainly the repetition of a section in the Act of 1859 (22 & 23 Vict. c. 56, s. 5). The former section, however, applied only to measures for exciseable liquors, but this has extended the provision to measures for all liquids. There are not many of these measures made.

Stamping of verification on weights for coin.
41 & 42 Vict. c. 49, s. 31.

Every coin weight, not less in weight than the weight of the lightest coin for the time being current, shall be verified and stamped by the Board of Trade with a mark of verification under the Weights and Measures Act, 1878, and otherwise shall not be deemed a just weight for determining the weight of gold and silver coin of the realm.

Every person who uses any weight declared by this section not to be a just weight shall be liable to a fine not exceeding fifty pounds.

Coin weight.—This is defined in the 70th section of the Act to mean "a weight used or intended to be used for weighing coin" (41 & 42 Vict. c. 49, s. 70). A similar provision for the stamping of coin weights was made in the 17th section of the Coinage Act, 1870, the greater portion of which section is now repealed.

The Warden of the Standards in his Eleventh Annual Report complains that the number of coin weights sent to the Board of Trade is very small, and in the Twelfth Report it is stated that the "provisions of the Coinage Act, which are now included in the Weights and Measures Act, 1878, which require all coin weights to be verified at the office of the Board of Trade, continue to be generally disregarded." Since the passing of the Coinage Act, 1870, it has always been a question whether the weights used in banks were subject to the provisions of that enactment. The bankers have contended that they use their weights, not to weigh the coin, but as a quick and easy method of counting it, and therefore they have not had their weights verified and stamped by the Board of Trade. This may account for the small number of weights stamped by the Board of Trade as noticed in the above-mentioned reports, for if the provisions with regard to coin weights only apply to those weights used for actually weighing the coin with a view to ascertain its value, the number of weights so used can not be very great. The Act of 1878 will scarcely clear up the matter, for although the use of a coin weight is not confined to a use for trade, the definition of a coin weight, as given above, will narrow the expression to a "weight used or intended to be used for weighing coin," and this can be interpreted according as a wide or restricted meaning is attached to the phrase "weighing coin."

The Orders in Council, dated August 9, 1870, legalising new standard coin weights, published in the *London Gazette* upon August 12, 1870, the *Edinburgh Gazette*, August 16, 1870, and the *Dublin Gazette*, August 16, 1870, in accordance with the provisions of the Coinage Act, 1870, and the Regulations of the Board of Trade relating to standard weights for coin made in pursuance of the same Act, and duly advertised in the same gazettes in April 1871, will still continue in force, although the Coinage Act, 1870, has now been repealed (41 & 42 Vict. c. 49, s. 86).

The above-mentioned Order and Regulations will both be found in the Appendix to the Fifth Annual Report of the Warden of the Standards, 1870–71.

A weight or measure duly stamped by an inspector under the Weights and Measures Act, 1878, shall be a legal weight or measure throughout the United Kingdom, unless found to be false or unjust, *Validity of weights and measures stamped throughout*

the United Kingdom.
41 & 42 Vict. c. 49, s. 45.

and shall not be liable to be re-stamped because used in any place other than that in which it was originally stamped.

This section reproduces in a somewhat different form the provisions of section 27 of the Weights and Measures Act, 1835 (5 & 6 Will. IV., c. 63, s. 27). It makes it clear, however, that this enactment applies only to weights and measures stamped by inspectors under the Weights and Measures Act, 1878.

Forgery, &c., of stamps on measures or weights.
41 & 42 Vict. c. 49, s. 32.

If any person forges or counterfeits any stamp used for the stamping under the Weights and Measures Act, 1878, of any measure or weight, or used before the commencement of that Act for the stamping of any measure or weight, under any enactment repealed by that Act, or wilfully increases or diminishes a weight so stamped, he shall be liable to a fine not exceeding fifty pounds.

Any person who knowingly uses, sells, utters, disposes of, or exposes for sale any measure or weight with such forged or counterfeit stamp thereon, or a weight so increased or diminished, shall be liable to a fine not exceeding ten pounds.

All measures and weights with any such forged or counterfeit stamp shall be forfeited.

The law as to forgery of stamps is here made to correspond with the law as to the forgery of stamps upon coin weights, contained in the 17th section of the Coinage Act, 1870 (33 & 34 Vict. c. 10, s. 17). As these provisions with regard to coin weights will be included in the section given above, the greater portion of the 17th section of the Coinage Act, 1870, has been repealed.

Wilfully increases or diminishes a weight so stamped.—There is here a most important and, it may be suggested, a most unfortunate omission, no mention being made of the wilful increasing or diminishing a stamped *measure*. It is difficult to discover the cause for this omission, unless it is an oversight, for experience teaches that measures are at least as likely to be tampered with as weights.

Shall be forfeited.—There is here, very properly, no discretion as to the forfeiture, and it will be the necessary result of a conviction under this section.

Many provisions also with regard to the stamping and verification of weights and measures will be found described and commented upon in the next chapter:—Inspectors and their Duties.

CHAPTER VIII.

INSPECTORS AND THEIR DUTIES.

Every local authority shall from time to time appoint a sufficient number of inspectors of weights and measures for safely keeping the local standards provided by such authority, and for the discharge of the other duties of inspectors under the Weights and Measures Act, 1878; and where they appoint more than one such inspector, shall allot to each inspector (subject to any arrangement made for a chief inspector or inspectors) a separate district, to be distinguished by some name, number, or mark; and the local authority may suspend or dismiss any inspector appointed by them or appoint additional inspectors, as occasion may require, and shall assign reasonable remuneration to each inspector for his duties. Appointment of inspectors of weights and measures. 41 & 42 Vict. c. 49, s. 43.

A local authority may, if they think fit, appoint different persons to be inspectors for verification and for inspection respectively of weights and measures under the said Act.

A maker or seller of weights or measures, or a person employed in the making or selling thereof, shall not be an inspector of weights and measures under the said Act.

An inspector of weights and measures shall forthwith on his appointment enter into a recognizance to the Crown (to be sued for in any court of record) in the sum of two hundred pounds for the due performance of the duties of his office, and for the due payment at the times fixed by the local authority appointing him, of all fees received by him under the Weights and Measures Act, 1878, and for the safety of the local standards and the stamps and appliances for verification committed to his charge, and for their due surrender immediately on his removal or other cessation from office to the person appointed by the local authority to receive them.

Every inspector already appointed in pursuance of any enactment repealed by the Weights and Measures Act, 1878, is to continue in office as if he had been appointed in pursuance of that Act (41 & 42 Vict. c. 49, s. 86).

Every local authority shall from time to time appoint.—The local authority is declared by the fourth schedule to the Weights and Measures Act, 1878, to be the Court of Quarter Sessions in counties, and the Town Councils in boroughs. The inspectors will therefore now be appointed either by the Courts of Quarter Sessions or the Town Councils, and the cases in which the appointment was vested in the Recorder will not occur. See *R.* v. *Recorder of Hull*, 8 A. & E. 638; 2 J. P. 550; and *Daly* v. *Sharood*, 3 Jur. N.S. 63; 20 J. P. 405 and 820; and Administration, *ante*.

Shall allot to each inspector a separate district.—The inspector will therefore be appointed for a special district, within which he must act, as, subject to the special power given to a county inspector under certain circumstances to act within another jurisdiction by the next section of the Act, each inspector will be appointed for the discharge of his particular duties within a certain district.

Different persons to be inspectors for verification and for inspection. —This permission to divide the duties of the inspectors seems to be a new one. In most cases the inspectors who verify the weights and measures will also carry out the duties of inspection. There may be some places where, owing to the extent of the area or density of the population, it may be advisable to separate the two duties, and this can now be done. In some divisions, where the inspectors have fixed days for stamping, advantage is sometimes taken of the knowledge that the inspector is otherwise engaged to use unjust weights, measures, and scales in his absence. This can now be prevented by using the provisions of the above section.

A maker or seller of weights or measures shall not be an inspector. —This clause, containing a provision which is obviously a most important one, has been so worded as to make it clear that an inspector may not sell or make weights or measures either before or after his appointment. The inspector should not adjust or mend a weight or measure in any way, his duties being confined to comparing, verifying, and stamping.

Shall forthwith enter into a recognizance.—This should be done as soon after the appointment as possible, for although an inspector might obtain his warrant and begin to act before doing so, there would be considerable risk attached to such a proceeding.

The local authority shall from time to time fix the times and places within their jurisdiction at which each inspector appointed by them is to attend for the purpose of the verification of weights and measures; and the inspector shall attend, with the local standards in his custody, at each time and place fixed, and shall examine every measure or weight which is of the same denomination as one of such standards, and is brought to him for the purpose of verification, and compare the same with that standard, and if he find the same correct shall stamp it with a stamp of verification in such manner as best to prevent fraud; and in the case of a measure or of a weight of a quarter of a pound or upwards, shall further stamp thereon a name, number, or mark distinguishing the district for which he acts.

<small>Verification and stamping by inspectors of weights and measures. 41 & 42 Vict. c. 49, s. 44.</small>

He shall also enter into a book kept by him minutes of every such verification, and give, if required, a certificate under his hand of every such stamping.

An inspector appointed by the local authority for a county may enter a place within the district of an inspector appointed by any other local authority, and there verify and stamp the weights and measures of any person residing within his own district, but if he knowingly stamp a weight or measure of any person residing in the district of an inspector legally appointed by another local authority, he shall be liable to a fine not exceeding twenty shillings for every weight or measure which he so stamps.

An inspector under the Weights and Measures Act, 1878, may take in respect of the verification and stamping of weights and measures such fees not exceeding those specified in the fifth schedule to that Act as the authority appointing him from time to time fix, and shall at such times not less often than once a quarter as the said authority direct, account for and pay over to the treasurer of the

<small>Fees for comparison and stamping. 41 & 42 Vict. c. 49, s. 47.</small>

local rate, or such person as the said authority direct, all fees taken by him.

Where the Board of Trade, upon the application of any local authority from time to time, represent to her Majesty that it would be expedient to alter the fees taken by the inspectors of such authority under the Weights and Measures Act, 1878 (whether specified in the said schedule to that Act or in any order previously made under this section), or for the purpose of adapting those fees to the local standards provided by such authority, to add to the said fees, it shall be lawful for her Majesty, by Order in Council, from time to time to alter or add to the said fees.

Shall examine every measure or weight which is of the same denomination as one of the standards.—It has already been shown in considering section 24 that all material weights and measures must now be of the same denomination as one of the Board of Trade standards. Following this enactment it is now made clear that an inspector cannot verify and stamp a weight or measure which is not of the same denomination as one of his standards; for example, that with a standard of a gallon he cannot verify a three gallon vessel. Hitherto the practice has not been uniform, one inspector stamping such a vessel, and another seizing it as illegal. As the inspector is bound to examine every measure or weight brought to him with the corresponding standard, it will be necessary for him to have a complete set of standards of all weights and measures which are now legal, that is, of all weights and measures which are contained in the second schedule of the Weights and Measures Act, 1878.

If he find the same correct shall stamp it.—It is important to note carefully the duties of the inspectors as required by this section, there having been much difference of opinion and diversity of practice in carrying out the provisions of the former Acts upon this subject. The inspector is to examine every measure or weight which is brought to him for the purpose, whether it has been stamped before or not. If it is incorrect there is an end of the proceeding, and the weight or measure must be returned with a notification to that effect. If it is correct, then the inspector must stamp it with a stamp of verification again whether it has been stamped before or not. The stamp is to be placed in such manner as best to prevent fraud, so that in cases where the plug of lead or pewter is inserted into a weight for the purpose of adjusting and stamping, the stamp should be affixed upon such plug so that as far as possible any tampering with the plug will necessarily lead to the injury and removal of the stamp.

Give if required a certificate.—A form for this certificate will be found in the *Appendix*.

All weights and measures which have been duly verified and stamped in pursuance of the Acts now repealed will continue, and be as valid as if they had been verified and stamped under the present Act, and that although such weights or measures could not now have been verified and stamped at all; and all weights and measures which before the first of January, 1879, might lawfully have been used without being stamped, but which now are required to be stamped, may be used without a stamp until the expiration of six months after the said date (41 & 42 Vict. c. 49, s. 86).

The last clause in section 44, given above, gives power to an inspector appointed by a county authority to enter into a district of an inspector appointed by another authority and there stamp the weights and measures of any person residing within his own district; but if he—the county inspector—knowingly stamps the weights or measures of a person living in the district of another inspector appointed by another local authority, he is to be liable to a fine of twenty shillings for every weight or measure which he so stamps. Each inspector being appointed to act within his own district cannot enter the district of another inspector to carry out the duties of the Weights and Measures Acts. When he leaves his district he ceases to be an inspector, and cannot, without distinct statutory authority, act as such. As a matter of convenience it often happened that the county inspectors' duties with regard to stamping could best be performed—with regard to the convenience both of the inspector and the public—within the limits of a county town, which was probably a borough with another jurisdiction. To meet this case the clause now under consideration was framed. It is a reproduction, with, however, some important alterations in the wording, of a somewhat similar clause in the Act of 1835 (5 & 6 Will. IV. c. 63, s. 25). The former clause empowered the county inspector to enter within the limits of another jurisdiction and there stamp the weights and measures of any person residing within his own district, but in creating the penalty it enacted that "*any inspector* knowingly stamping any weight or measure of any person residing within the limits of any local jurisdiction for which another inspector may have been legally appointed shall forfeit a sum &c." It was held that the words "any inspector" here used applied to inspectors appointed in boroughs as well as in counties, and that this was a general prohibition against any inspector knowingly stamping the weight or measure of any person residing within the district of another inspector (*R.* v. *Skelton*, 1 E. & E. 816; 28 L. J. M. C. 222; 33 L. T. 120; 7 W. R. 447; 5 Jur. N.S. 1347; 23 J. P. 630). In this decision the words used were very wide indeed. The object of the clause was stated to be to prevent unseemly conflict between county and borough inspectors, but in the judgment delivered the prohibition was said to extend to any inspector knowingly stamping the weights or measures of persons residing within the district of another inspector. It will be noticed that the clause in the new Act restricts the prohibition to the county inspector who, acting under the powers given him by the said clause, abuses the authority thus given, and knowingly stamps the weights or measures of persons residing within the limits he has been permitted to enter. And this is really the only prohibition that is required, for no other inspector under any other circumstances than those provided for under this section will have the opportunity of committing the offence here described. If any other inspector in any other way were to act without due authority there would be ample legal provision, civil and criminal, to provide for or punish such a proceeding.

Fees for comparison and stamping.—The local authority is to fix the scale of fees to be taken by the inspectors for stamping weights and measures, but the fees are not to exceed in amount those specified in the fifth schedule of the Act of 1878, which will be found in the *Appendix*. The fees are to be taken for verifying and stamping the weights and measures. Where, therefore, the weight or measure is found to be incorrect, and is consequently returned, there will be no fee received; but in all other cases the weight or measure being found correct will be stamped and the fee charged. The fifth schedule to the Act of 1878 will be found to contain fees for the stamping of nearly all the weights and measures for which there are standards, and it is these only that the inspector can verify and stamp. A new power is given in the present Act to the Queen in Council, upon application from the local authority, to adopt the schedule of maximum fees to any altered circumstances that may arise; and as there is at present no provision made for measures of length except the yard, and for the yard only when it is made of wood, an early revision of the present table would seem to be advisable.

Power to inspect measures, weights, scales, &c., and to enter shops, &c. for that purpose. 41 & 42 Vict. c. 49, s. 48.

Every inspector under the Weights and Measures Act, 1878, authorised in writing under the hand of a justice of the peace, also every justice of the peace, may at all reasonable times inspect all weights, measures, scales, balances, steelyards, and weighing machines within his jurisdiction which are used or in the possession of any person, or on any premises for use for trade, and may compare every such weight and measure with some local standard, and may seize and detain any weight, measure, scale, balance, or steelyard, which is liable to be forfeited in pursuance of the said Act, and may, for the purpose of such inspection, enter any place, whether a building or in the open air, whether open or enclosed, where he has reasonable cause to believe that there is any weight, measure, scale, balance, steelyard, or weighing machine which he is authorised by the said Act to inspect.

Any person who neglects or refuses to produce

for such inspection all weights, measures, scales, balances, steelyards, and weighing machines in his possession or on his premises, or refuses to permit the justice or inspector to examine the same or any of them, or obstructs the entry of the justice or inspector under this section, or otherwise obstructs or hinders a justice or inspector acting under this section, shall be liable to a fine not exceeding five, or, in the case of a second offence, ten pounds.

Authorised in writing under the hand of a justice of the peace.—A general warrant will be sufficient; there is no necessity for the inspector to have a special warrant in each individual case (*Hutchins* v. *Reeves*, 9 M. & W. 747; 11 L. J. M. C. 109; 6 Jur. 439; 6 J. P. 313). A justice can only give this authority to an inspector appointed under the Weights and Measures Act, 1878; nor can an inspector depute anyone to perform his duties for him. Forms of this warrant, which may be used, will be found in the *Appendix*. Upon the death of the justice signing the same, or upon his ceasing to act, a fresh warrant should be obtained.

Which are used or in the possession of any person, or on any premises for use for trade.—The powers given to an inspector by the present section are more general than those hitherto possessed by him under the former enactments. By the 28th section of the Act of 1835, the inspector might enter "any shop, store, warehouse, stall, yard, or place whatsoever within his jurisdiction, wherein goods were exposed or kept for sale, or were weighed for conveyance or carriage"; and in the Act of 1859, by section 3, he was also empowered " to inspect all beams, scales, and balances, and weights and measures, in the possession of any person selling, offering, or exposing for sale any goods on any open ground, or in any public street, lane, thoroughfare, or other open place" (5 & 6 Will. IV. c. 63, s. 28; 22 & 23 Vict. c. 56, s. 3). In many cases doubts arose whether an inspector had a right to enter premises in which he found unjust weights and measures which he had power to seize, and for which the possessor was fined (See *Kershaw* v. *Johnson*, 1 C. & K. 329). Whether a market hall was an " open place " under the latter section, or whether a farmhouse where butter was sometimes sold, was a " place wherein goods were exposed or left for sale " under the former section, were points upon which difference of opinion has been expressed. A barn or outhouse was held to to be possibly a " place, &c.," if there was evidence of using weights within it; but where there was no evidence of an actual sale having taken place in the barn, it was deemed not to be within the provisions of the 28th section of the Act of 1835 (*Griffiths* v. *Place*, 33 J. P. 629).

Now, however, by the 48th section of the Act of 1878, given above, a general power is given to an inspector to inspect all weights, measures, scales, balances, steelyards, and weighing machines used or kept for use for trade, and to seize any weight, measure, scale, balance, or steelyard, which is liable to be forfeited, and to enter premises where the same may be found; so that the power of entry is extended to every case where there is a weight, measure, &c., used in trade, and subject, if unauthorised or unjust, to be forfeited. The definition of "trade" has already been given in section 19 of the same Act. It includes "every contract, bargain, sale, or dealing made or had in the United Kingdom for any work, goods, wares, or merchandise, or other thing which has been or is to be done, sold, delivered, carried or agreed for by weight or measure, and all tolls and duties charged or collected according to weight or measure" (41 & 42 Vict. c. 49, s. 19). Where, also, any weight, measure, scale, &c., is found in the possession of any person carrying on trade within the meaning of the above definition, or on the premises of any person which, whether a building or in the open air, whether open or enclosed, are used for trade, within the same meaning, such person shall be deemed, until the contrary is proved, to have such weight, measure, scale, &c., in his possession for use for trade (41 & 42 Vict. c. 49, s. 59). See Legal Proceedings, *post*.

May compare.—The inspector should not seize the weight, measure, scale, &c., without having first compared it with the standards in order to ascertain whether it was just or not (*Kershaw* v. *Johnson*, 1 C. & K. 329).

May seize and detain.—Hitherto there has been no power to seize a scale, balance, or steelyard, except where it was being used in the streets (*Thomas* v. *Stephenson*, 2 E. & B. 108; 22 L. J. Q. B. 258; 17 Jur. 597; 1 W. R. 325; 17 J. P. 537). Now the inspector can seize the scales as well as the weights and measures in all cases where they are liable to be forfeited. The seizure of an unjust scale will, however, probably only be carried out in extreme cases, where the scale is either worn out, or is so fraudulently or badly constructed as to be a source of continual trouble to the possessor, and mischief to the public. There are some scales which can be made to weigh justly or unjustly according to the will of the person using them, by placing the weights in different positions, or by other means known to those experienced in such practices. The new power now given to the inspector will exactly meet such cases, and enable the authorities to stop at once such extensive systems of fraud. There is no power given to seize a "weighing machine" as the term is here applied, in its technical sense, to mean the large fixed machines which the inspectors would not be able to remove.

The last clause in this section provides a penalty for interfering with or obstructing an inspector in the execution of his duties. It applies only to inspectors appointed under the Weights and Measures Act, 1878, and to them only when they are in the execution of their duties under that Act. The first two offences mentioned, "neglecting or refusing to produce," and "refusing to permit examination," will apply only to persons who have the weights, measures, scales, &c., in their possession, or upon their premises; but the other offences of "obstructing the entry," and "otherwise obstructing or hindering" will apply generally to any manager, shopman, or other person who may commit them. This clause has been made co-extensive with the rest of the section, and by that means supplies an important omission in the former Acts which provided no penalty for obstructing an inspector when examining or seizing a weight, measure, scale, &c., which was being used in the streets. The increased fine in the case of a second offence, committed within five years of a previous conviction of an offence under the same section, is also introduced for the first time in the Act of 1878.

If an inspector under the Weights and Measures Act, 1878, stamps a weight or measure in contravention of any provision of that Act, or without duly verifying the same by comparison with a local standard, or is guilty of a breach of any duty imposed on him by that Act, or otherwise misconducts himself in the execution of his office, he shall be liable to a fine not exceeding five pounds for each offence. Penalty on inspector for misconduct. 41 & 42 Vict. c. 49, s. 49.

This is substantially a re-enactment of a similar provision in the Act of 1835 (5 & 6 Will. IV. c. 63, s 29). A doubt, however, whether that section applied to any person other than the one now called an inspector is removed, as the present enactment distinctly applies only to "inspectors under the Weights and Measures Act, 1878"

In addition to this liability, a county inspector who enters the district of an inspector appointed by another local authority under the powers given to him by section 44 of the Act of 1878, and when there knowingly stamps a weight or measure belonging to a person not residing within his district, is liable to a fine not exceeding twenty shillings for every weight or measure so stamped (41 & 42 Vict. c. 49, s. 44).

Superintendents and inspectors of rural police under 2 & 3 Vict., c. 93. may be made inspectors of weights and measures; but they should not be paid a separate salary as such. See 2 & 3 Vict. c. 93, s. 10; and R. v. Jarvis, 3 E. & B. 640.

CHAPTER IX.

LEGAL PROCEEDINGS.

(1.) *Summary Proceedings.*

Prosecution of offences and recovery of fines.
41 & 42 Vict. c. 49, s. 56.

All offences under the Weights and Measures Act, 1878, may be prosecuted, and all fines and forfeitures under that Act may be recovered on summary conviction before a court of summary jurisdiction in manner provided by the Summary Jurisdiction Act.

The Court, when hearing and determining an information or complaint under the Weights and Measures Act, 1878, shall be constituted either of two or more justices of the peace in petty sessions, sitting at a place appointed for holding petty sessions, or of some magistrate or officer sitting alone or with others, at some Court or other place appointed for the administration of justice, and for the time being empowered by law to do alone any act authorised to be done by more than one justice of the peace.

In the present Act the modern clauses for summary legal proceedings have been substituted for the old ones, and the procedure generally is made consistent with the present practice.

Court of Summary Jurisdiction.—This is defined in the 70th section of the Act, to mean " any justice or justices of the peace, metropolitan police magistrate, stipendiary or other magistrate or officers, by whatever name called, to whom jurisdiction is given by the Summary Jurisdiction Act or any Acts therein referred to."

In manner provided by the Summary Jurisdiction Act.—The "Summary Jurisdiction Act" means the "11 & 12 Vict. c. 43, inclusive of any Acts amending the same;" that being the Act which regulates the performance of the duties of justices of the peace out of sessions within England and Wales with respect to summary convictions and orders. It is one of the Acts popularly known as Jervis's Acts, and it is the statute by which the proceedings of justices under their summary jurisdiction are governed. For full and complete descriptions of the various provisions of this Act, together with exhaustive explanations of the points of practice arising in bringing the Act into effect, it will be necessary to refer to one of the numerous Justices' Manuals which have been published. It is only proposed here to give a short abstract of the principal provisions of the statute, and to add some suggestions which may be of help in applying the Act to the punishment of offences under the Weights and Measures Act.

It should, however, be noted that a Bill to amend the law with regard to the summary jurisdiction of magistrates is about to be introduced into Parliament, so that considerable alterations may shortly be made in the procedure before justices.

The information.—This must be laid within six calendar months from the time when the offence was committed It need not be upon oath, nor need it be in writing. In some divisions, however, it is customary for it to be in writing, and when this is the case it should state the name and occupation of the person charged, and the offence, together with the time and place when and where the offence was committed. The description of the offence may be in the words of the section creating it, or in similar words; and no objection is to be allowed to any information for any alleged defect in substance or form. The offences under the Weights and Measures Act, not being matters of individual grievance but of public policy and utility, any person has a general power to inform and sue for the penalties; but the Inspector will nearly always be the informer, as he is in most cases the only person who has the power to discover the offence and prove that it has been committed. The information must be for one offence only, as to which see the remarks upon the 25th section of the Weights and Measures Act. 1878, in the chapter upon Unjust Weights, Measures, and Weighing Machines, *ante*. Several offenders may be joined in the same information and convicted in separate penalties, where the offence complained of admits of the participation of several persons.

The summons.—This, like the information, must shortly describe the offence, and in so doing it will be sufficient to use the words of the section of the Act creating the offence for which the person is summoned. One justice may receive the information and grant a summons and do all the necessary acts preliminary to the hearing, although the summons must be heard by two or more justices; and after the hearing one justice may issue all warrants of distress and commitment. The summons must be served by a constable or other peace officer or other person to whom it has been delivered upon the party to whom it is directed, either personally or by leaving it with some person for him at his last or most usual place of abode. It should be served a reasonable time before the hearing, and the justices are to decide what is a reasonable time, having regard to the circumstances of each case.

The warrant.—If the person summoned does not appear, the justices, upon proof of the due service of the summons, have power

F

to hear the case in the defendant's absence. But where practicable they should be satisfied that the summons has not only been served but has actually been brought to the defendant's notice. Under the provisions of the present Weights and Measures Act it will be advisable to enforce the attendance of the defendants in all cases, and neither hear the case in their absence, or allow their wives or managers to appear for them. Under the former Acts both these courses have often been pursued, and it may thus have happened that persons may have been convicted and the penalties paid for them without their knowledge. As, however, it will now be necessary to provide for the easy proof of a previous conviction, and as it is in all cases more convenient to have the person summoned actually before the bench at the hearing, and especially so when an immediate committment is to follow the non-payment of the penalty, the attendance of the defendant should be enforced. This may be done by a second summons, or by issuing a *warrant* to apprehend the defendant. This warrant may be issued by the justices after proof by oath or affirmation of the service of the summons a reasonable time before the time appointed for the hearing, and after receiving upon oath or affirmation evidence substantiating the matter of the information. The warrant must be under the hand and seal or hands and seals of the justice or justices issuing the same; it should state shortly the matter of the information, should name or otherwise describe the offender, and should order his apprehension. The warrant need not be made returnable at any particular time, but may remain in force until it is executed; and no objection to the warrant is to be allowed for any alleged defect in substance or form.

The hearing.—This must be by two or more justices in petty sessions assembled, sitting in an open and public court, to which the public may have access as far as it can conveniently contain them. The person charged must be admitted to make his full answer and defence, and to have the witnesses examined and cross-examined by counsel or attorney; and every informant is at liberty to conduct the information and to have the witnesses examined and cross-examined by counsel or attorney. If the defendant appears at the hearing, he must then object to the form of the summons, or to any irregularity in the proceedings on which it was founded. If he does not, but asks for judgment in his favour or upon the merits, he waives any irregularity in the process for bringing him before the court, his appearance curing any irregularity in the service of the summons, or the want of one. If before the time for the hearing the inspector dies, the proceedings will be at an end; or if the inspector does not attend by himself, his counsel or attorney, the justices can dismiss the information, or they may adjourn the hearing to some future day upon such terms as they may think fit. The defendant may appear by attorney, but if the justices deem the personal appearance of the party charged to be necessary, they have power to enforce it.

Both parties being before the justices, the substance of the information should be stated to the defendant, and if he admits the truth of the same, the justices present can convict him, or make an order accordingly. The justices have full discretion in the matter, but it is advisable to prove shortly the offence, as stated in the summons, even where the facts are admitted. The inspector may now give evidence, irrespective of any pecuniary interest he may have in the penalty or result. Before 1851 this was not so, and the offences under the Weights and Measures Acts were proved by means of assistants who used to accompany the inspector upon his rounds, and being present when he examined the weights, measures, and scales, they were able to prove the facts of the case. This system is still in force in many divisions, although for the last twenty-seven years the inspectors might have been examined in support of the summons. It should be remembered that the principal question which the justices have to determine in the case of unjust weights, measures, and scales, is the condition of the same *at the time* they were examined by the inspector. As the defendant cannot give evidence upon his own behalf, and cannot call his wife as a witness, the only method of successfully disputing the inspector's decision is to call in some independent witness at the time of the examination, who can then be called for the defence at the time of the hearing. The justice having heard the inspector and such witnesses as he may examine, and such other evidence as he may adduce in support of his information, and having heard the defendant and such witnesses as he may examine, and such other evidence as he may adduce in his defence, and having also heard such witnesses as the inspector may examine in reply, will then consider the whole matter and determine the same, and convict or dismiss the information as the case may be. Neither the inspector or the defendant may address any observations in reply upon the evidence given by the other side. Before or during the hearing, the justices may adjourn the case to a certain time and place then appointed and stated in the presence and hearing of the parties or their attorneys or agents present; and they may suffer the defendant to go at large, or may commit him to such safe custody as they shall think fit, or may discharge him upon his entering into a recognizance, with or without sureties, conditioned for his appearance at the time and place to which the hearing is adjourned. If at the adjournment either or both of the parties do not appear, the justices may proceed with the hearing in their absence, or if the inspector does not appear, may dismiss the information with or without costs.

In all cases of summary conviction, the justices may, at discretion, award and order that the defendant shall pay to the inspector such costs as to the justices shall seem reasonable; and in cases where the summons is dismissed, the justices may in the same manner, at discretion, award and order that the inspector shall pay to the defendant such costs as to the justices shall seem just and reasonable. The amount of the costs must be fixed by the justices themselves, and stated in the conviction or order of dismissal; and they can be recovered in the same manner and under the same warrants as the penalty. If there is no penalty, they can be recovered by distress, or in default of distress, by imprisonment with or without hard labour, for any time not exceeding one calendar month, unless such costs are sooner paid.

The justices have power to summon witnesses to attend and give evidence, and to issue a warrant if the summons is disobeyed. Persons aiding and abetting in the commission of an offence are liable to be proceeded against and convicted for the same, either together with the principal offender, or before or after his conviction, and are liable upon conviction to the same forfeiture and punishment.

The conviction.—Here also, the offence may be described shortly in the words of the Act or in similar words. If costs have been granted, the amount should be stated in the conviction, which should also contain the forfeiture of any weights, measures, or scales, which are to be forfeited. The formal conviction may be drawn up at any time before the return of a *certiorari*, although, after a commitment, or after the penalty has been levied by distress, or after action brought against the magistrates. But it should be drawn up and lodged with the clerk of the peace, to be by him filed among the records of the court of Quarter Sessions. This is the more important now that certified copies of the conviction will be wanted for evidence in the case of second offences under the same section of the Act within five years. See *post*. The conviction need only be signed by two justices, although more may have been present at the decision; but the two justices signing the conviction must have been present, and acted together during the whole of the hearing and determination of the case. The conviction should state the manner in which the penalty is to be applied now that the justices have a discretion in ordering a moiety of the fine to be paid to the informer.

The distress warrant.—Upon the non-payment of a penalty or the costs, the justices convicting, or any justice of the same county, borough, or place, may issue a warrant of distress for levying the same. When, however, it may appear to any justice to whom application is made for such a warrant, that the issuing thereof would be ruinous to the defendant or his family, or whenever it shall appear by the confession of the defendant or otherwise, that he has no goods or chattels whereon to levy the distress, the justice may, instead of issuing a warrant of distress, commit the defendant to prison in such manner as by law he might have been committed if such warrant of distress had issued, and no goods had been found. When issuing a distress warrant, the justice may suffer the defendant to go at large or may order the defendant to be kept in safe custody until the return be made to the warrant, unless he give sufficient security, by recognizance or otherwise, to the satisfaction of the justice, for his appearance before him at the time and place appointed for the return of the warrant. If the goods of a person are not sufficient to satisfy the whole of the levy, they should not be taken, but it should be returned that no sufficient distress can be found; and if a married woman is fined, there can be no levy upon the goods of her husband to pay the penalty.

The commitment.—In case it be returned to a warrant of distress that no sufficient goods can be found, the justice to whom such return is made, or any other justice of the same county, borough, or place, may, by warrant, commit the defendant to prison for any term not exceeding three calendar months, unless the sum adjudged to be paid, and all costs of the distress, commitment, and of conveying the defendant to prison, shall be sooner paid. The amount of the costs must be ascertained and stated in the commitment; and it should be noted that there is no power to award hard labour.

The foregoing provisions will apply to all cases under the Weights and Measures Act, 1878, where there has been a previous conviction, and where the penalty may consequently be above five pounds. In the case of first offences, and where the penalty is less than five pounds, the proceedings will be governed by the provisions of the Small Penalties Act, 1865 (28 & 29 Vict. c. 127). This Act provides that when upon summary conviction any offender may be adjudged to pay a penalty not exceeding five pounds, such offender, in case of non-payment thereof, may, without any warrant of distress, be committed to prison for any term not exceeding the period specified in the following scale, unless the penalty shall be sooner paid: For any penalty not exceeding ten shillings, the imprisonment not to exceed seven days; exceeding ten shillings and not exceeding one pound, fourteen days; exceeding one pound but not exceeding two pounds, one month; exceeding two pounds but not exceeding five pounds, two months. There is in this Act no power to award hard labour in cases under the Weights and Measures Act; and where the penalty—including costs—does not exceed five pounds, the scale above given takes the place of the more general provision, which applies only to cases where the penalty and costs exceed five pounds. But the justices have still the discretion of issuing a distress warrant before commitment, if they should think it advisable so to do.

The justices have power to state a " case " for the opinion of one of the superior courts of law, if either party is dissatisfied with the justices' decision as being erroneous in point of law. This power is given to them by the Act of 20 & 21 Vict. c. 43, which provides that either party to the proceeding before the justices may, if dissatisfied with their determination as being erroneous in point of law, apply in writing within three days after the same to the justices, to state and sign a case setting forth the facts and grounds for such a determination, for the opinion thereon of one of the superior courts of law, to be named by the party applying; and such party, called the "appellant," shall, within three days after receiving such case, transmit the same to the court named in his application, first giving notice in writing of such appeal, with a copy of the case so stated and signed, to the other party to the proceeding in which the determination was given, who is called the "respondent." The justices may refuse to state a case if they are of opinion that the application is merely frivolous, but not otherwise; and the Queen's Bench Division may compel them to state a case if they refuse to do so. The court to which a case is transmitted may reverse, affirm, or amend the determination in respect of which the case has been stated, or may remit the matter to the justices with the opinion of the court thereon, or may make such other order in relation to the matter as to the court may seem fit; and all such orders shall be final and conclusive on all parties. There are various other provisions with regard to giving notices, entering into recognizances, filing affidavits, amending the case, &c., which will all be found fully described and commented upon both in Oke's "Magisterial Synopsis" and Stone's "Justices' Manual."

Provisions as to summary proceedings. 41 & 42 Vict. c. 49, s. 57.

The following enactments shall apply to proceedings under the Weights and Measures Act, 1878, before a court of summary jurisdiction; that is to say,

1. The description of any offence in the words of that Act, or in similar words, shall be sufficient in law; and
2. Any exception, exemption, proviso, excuse, or qualification, whether it does or does not accompany in the same section the description of the offence, may be proved by the defendant, but need not be specified or negatived in the information or complaint, and, if so specified or negatived, no proof in relation to the matter so specified or negatived shall be required on the part of the informant or complainant; and
3. A warrant of commitment shall not be held void by reason of any defect therein, if it be therein alleged that the offender has been convicted, and there is a good and valid conviction to sustain the same.
4. Such portion of any fine under the said Act, not exceeding a moiety, as the court of summary jurisdiction before whom a person is convicted think fit to direct, may, if the Court in their discretion so order, be paid to the informer.
5. All weights, measures, scales, balances, and steelyards forfeited under the said Act shall be broken up, and the materials thereof may

be sold or otherwise disposed of as a court of summary jurisdiction direct, and the proceeds of such sale shall be applied in like manner as fines under the said Act.

A person shall not be liable to any increased penalty for a second offence under any section of the Weights and Measures Act, 1878, unless that offence was committed after a conviction within five years previously for an offence under the same section. *Limitation as to conviction for second offences. 41 & 42 Vict. c. 49, s. 58.*

Where any weight, measure, scale, balance, steelyard, or weighing machine is found in the possession of any person carrying on trade within the meaning of the Weights and Measures Act, 1878, or on the premises of any person which, whether a building or in the open air, whether open or enclosed, are used for trade within the meaning of that Act, such person shall be deemed for the purposes of that Act, until the contrary is proved, to have such weight, measure, scale, balance, steelyard, or weighing machine in his possession for use for trade. *Evidence as to possession. 41 & 42 Vict. c. 49, s. 59.*

A portion of any fine not exceeding a moiety may be paid to the informer.—The power to grant a portion of the fine, not exceeding a moiety, to the informer, has been expressly retained in the Weights and Measures Act, although not consistent with the modern English practice. Under the Act of 1835, there was no discretion given to the justices in the matter, except as to the amount to be deducted (5 & 6 Will. IV. c. 63, s. 32). Now the whole question is left in the hands of the justices. The retention of this almost obsolete custom is a matter about which there is much difference of opinion. On the one hand it is argued that there being no real supervision over the inspectors, the only method of keeping them up to their work is to give them a pecuniary interest in the number of convictions they obtain; on the other side it is contended that the system of "blood money," as it is called, places the inspector in a most unpleasant position, and giving him an interest in the commission of offences rather than in their suppression, lays him open to temptations to which he should not be exposed. The inspectors as a rule would much prefer to be paid entirely by salary, and if they are fit to be entrusted at all with the very important duties allotted to them they might, it is suggested, be safely left to conscientiously discharge the same with assiduity and attention. The general public also would have more confidence in the pure and strict administration of justice if the prosecution of offences were left in the hands of persons who were absolutely independent of the result of the proceedings.

The proceeds of such sale shall be applied in like manner as fines under this Act.—This will give the justices power, if they should order a sale, to grant a portion of the proceeds of such sale, not exceeding a moiety, to the inspectors. Under the former Acts the whole of the proceeds of such sale went to the County Treasurer (37 Geo. III. c. 143, s. 3; 55 Geo. III. c. 43, s. 2).

Conviction for second offences.—The second offence must have been committed within five years of the time of the former conviction. This former conviction also must have been for an offence under the same section of the Act of 1878. So that for the purpose of extending the limits of the penalty, according to the provisions of the Act of 1878, convictions under the former Acts cannot be used. These old convictions may still, however, be brought to the notice of the justices for their information, after they have determined to convict in the case before them. In proceeding against a person as for a second offence, it should be stated in the information and summons that he has been previously convicted for an offence under the same section, within five years previously to the committal of the second offence. In charging the defendant with the second offence, the previous conviction should not be mentioned, but the second offence should be determined first. When the justices have decided to convict upon the second offence, the defendant should be charged with having, within five years of the time of committing the second offence, been previously convicted of an offence under the same section. If the defendant admits the previous conviction the justices can then decide upon the whole case, and may inflict the increased penalty; but if the defendant denies the previous conviction it must be proved by legal evidence. This can be done, in accordance with the provisions of the Prevention of Crimes Act, 1871, by the inspector producing a certified copy of the conviction duly signed by the clerk of the peace or his deputy, and proving upon oath that the person so convicted is the same person who has just been convicted of the second offence (34 & 35 Vict. c. 112, s. 18). If the former conviction has not yet been returned to the clerk of the peace, a copy of the conviction, duly certified and signed by the clerk to the justices, will be sufficient. The inspector will have to pay a fee of five shillings for the copy of the conviction, but this fee can be allowed by the justices in awarding costs to the inspector.

Evidence as to possession.—This section is a new one, and is inserted in consequence of the alteration made in the law which renders the mere possession of an unjust weight, measure, or scale no offence, unless it is in possession for use for trade. Such a limitation would make it a difficult matter to convict if the burden of proof had not been thrown upon the defendant, who will have to show to the satisfaction of the justices that the weight, measure, or scale, &c., was not in his possession for use for trade. The provisions of this section have already been noticed in several places, wherever in fact the words "possession for use for trade" have been used.

The provision of section 36 of 5 & 6 Will. IV. c. 63, taking away the *certiorari* and prohibiting the quashing of a proceeding for want of form, has been omitted in the Act of 1878. Such provisions are not required under modern procedure; and whenever the writ of *certiorari* is taken away by statute, the Quarter Sessions have no power to state a special case for the opinion of the Queen's Bench, unless with the express consent of both parties. See *post*.

(2.) *Proceedings upon Appeal.*

Any person who feels himself aggrieved by a conviction or order of a court of summary jurisdiction under the Weights and Measures Act, 1878, may appeal therefrom, subject in England to the conditions following; that is to say, Appeal from conviction. 41 & 42 Vict. c. 49, s. 60.

1. The appeal shall be made to the next practicable court of general or quarter sessions having jurisdiction in the county or place in which the decision of the court was given, and holden not less than twenty-one days after the day on which such decision was given; and
2. The appellant shall, within ten days after the day on which the decision was given, serve notice on the other party and on the clerk of the court of summary jurisdiction of his intention to appeal, and of the general grounds of such appeal; and
3. The appellant shall, within three days after the day on which he gave notice of appeal, enter into a recognizance before a court of summary jurisdiction, with or without a surety or sureties as the court may direct, conditioned to appear at the said sessions and to try such appeal, and to abide the judgment of the court thereon, and to pay such costs as may be awarded by the court; or the appellant may, if the court of summary jurisdiction thinks it expedient, instead of enter-

ing into a recognizance, give such other security, by deposit of money with the clerk of the court of summary jurisdiction or otherwise, as the court deems sufficient ; and

4. Where the appellant is in custody, a court of summary jurisdiction may, if it seem fit, on the appellant entering into such recognizance, or giving such other security as aforesaid, release him from custody; and

5. The court of appeal may adjourn the hearing of the appeal, and upon the hearing thereof may confirm, reverse, or modify the decision of the court of summary jurisdiction, or remit the matter to the court of summary jurisdiction, with the opinion of the court of appeal thereon, or make such other order in the matter as the court thinks just. The court of appeal may also make such order as to costs to be paid by either party as the court thinks just ; and

6. Whenever a decision is reversed by the court of appeal, the clerk of the peace shall indorse on the conviction or order appealed against a memorandum that such conviction or order has been quashed, and whenever any copy or certificate of such conviction or order is made, a copy of such memorandum shall be added thereto, and shall be sufficient evidence that the conviction or order has been quashed in every case where such copy or certificate would be sufficient evidence of such conviction or order ; and

7. Every notice in writing required by this section to be given by an appellant may be signed by him, or by his agent on his behalf, and may be transmitted in a registered letter by the post in the ordinary way, and shall be

deemed to have been served at the time when it would be delivered in the ordinary course of the post.

This section takes the place of one of a similar character in the Act of 1835 (5 & 6 Will. IV. c. 63, s. 35). The provisions have been altered to suit the requirements of modern practice, the section being for the most part the usual one now inserted where an appeal to Quarter Sessions is given. It is similar in most of its details to the usual form which will continue to be used until the time, which it is to be hoped is not far distant, when the procedure with regard to all appeals to Quarter Sessions will be made uniform

For further particulars with regard to appeals to Quarter Sessions it will be necessary to consult one of the many books which have been published upon the jurisdiction and practice of the Court of Quarter Sessions. The following general information may, however, be useful to persons using this book.

The appeal.—There can be no appeal against a summary conviction, unless expressly given by the statute under which the conviction takes place. The appeal is given by the Act to "any person who feels himself aggrieved by a conviction or order of a court of summary jurisdiction;" and these words must be taken to mean a person who has some special personal grievance, and not to include any captious person who may choose to take proceedings. The grievance also must be immediate and not consequential, and it must be a reasonable and substantial one. The appeal must be made to the "next practicable" court of general or quarter sessions; and where by the practice of the sessions or by statute notice of appeal is required, and there is not time before the next sessions to give the requisite notice, these are not the "next practicable" sessions. If one of the days intervening between the conviction and the sessions be a Sunday, the courts will exclude that day in judging whether the next sessions were practicable or not; and it is the first day of the sessions which is or is not practicable, for the sessions are always considered in law as one day, to whatever period they may by accidental causes be extended. The appeal is to be to the sessions of the jurisdiction in which the conviction was made, and holden not less than twenty-one days after "the day on which the decision was given."

The notice of appeal.—This notice may be signed by the appellant or by his agent, and may be sent through the post in a registered letter, it being deemed to have been served at the time when it would be delivered in the ordinary course of the post. It must be served upon the respondent, and upon the justices' clerk, and must contain the general grounds of the appeal. The directions of the Act must be strictly complied with, and the giving notice, when required, is a condition precedent to the right of appeal, and no practice of sessions which differs from these provisions can be allowed to prevail. The notice should be directed to the persons to whom it is required to be given, by their names and additions, and it should state that the appellant "feels himself aggrieved" by the conviction or order against which he appeals. The object of the notice being to make the other side aware of the intention to appeal against a particular conviction or order, if, in describing it, reasonable information is given by which they cannot be misled, the court will not construe that notice with the accuracy of a pleading. The notice is to contain the general grounds of the appeal; and a ground of appeal against a conviction stating that the appellant is "not guilty of the said offence," is sufficient to put in issue all the matters, without the concurrence of which the offence could not have been committed. It is important to note that the appellant will be confined to the grounds of appeal specified by him in his notice; and wherever any statement is not denied by the grounds of appeal, the respondent is not bound to prove it. Where grounds of appeal are frivolous, the court may order the appellant to pay the whole or part of the costs incurred by the respondent in disputing such grounds. A notice of appeal, once given, may be abandoned, and a fresh one given, if the time for appealing and giving notice be not expired; but if the notice is once given, the Sessions may, upon proof of notice of appeal, though no appeal was prescribed or entered, give such costs as they may think reasonable to the persons receiving such notice.

The recognizance.—The requirements of this clause must be strictly complied with, as the entering into a recognizance or the deposit of money is a condition precedent to the right to appeal. A married woman cannot enter into a recognizance herself, but she may do so by her husband as surety for her; and an infant ought to be bound by sureties.

The entry of an appeal.—The entry of the appeal with the clerk of the peace must be done within the time limited by the practice of the sessions. When entered, the court may adjourn the hearing of the appeal from one session to another.

The hearing.—This should take place in the presence of the appellant who has entered into a recognizance to *appear* at the sessions and try the appeal. The first proceeding is for the appellant to prove that he has complied with all the conditions enforced by the statute giving the appeal as conditions precedent, either to the right to appeal, or to the right to be heard upon the appeal. If he fails to prove such conditions as affect the right to appeal, the appeal must be dismissed; but if there be a failure of proof of conditions which affect the right to be heard, the court may, if they think fit, adjourn the appeal, if the defect be such as will admit of being remedied. The requisite conditions of appeal may, however, be admitted by the respondent, or the proof of them may be waived. When the appeal is against a conviction, the clerk of the peace reads the record of conviction returned to the sessions by the convicting justices, and this record is the only one of which the sessions can take notice. If the conviction has not been returned to the sessions, a *subpœna duces tecum* should be served upon the clerk to the justices, by whom it was made. The court has certain powers to amend any defects of form in the conviction; and any application to amend, or any objection to the grounds of appeal, &c., should be taken at this stage. The respondent will then begin and prove his case in support of the conviction appealed against, which being done, the appellant will either contend that the case has not been proved, or seek to rebut it by evidence. Neither the appellant nor respondent is confined to the evidence given before the convicting justices. The court will then have power to "confirm, reverse, or modify the decision of the justices, or remit the matter to them with the opinion of the court of appeal thereon, or may make such other order in the matter as they may think just."

The costs of the appeal.—The court of appeal may make such order as to costs, to be paid by either party, as it thinks just. The justices in the court below are not obliged to appear to support their conviction, and if they are made respondents, and an order is made upon them by Quarter Sessions to pay costs, the order will be quashed by the Queen's Bench, with costs. In an appeal against a conviction, the informant or defendant are the "parties" liable to pay costs, although the justices may be formal parties to the appeal. The order awarding costs must direct that such costs be paid to the clerk of the peace, to be by him paid over to the party entitled to them, and it should state within what time such costs are to be paid; and if the same are not paid within the time limited, and the party ordered to pay them is not bound by recognizance conditioned to pay such costs, the clerk of the peace will, upon application of the person entitled to the costs, or someone on his behalf, and on the payment of a fee of one shilling, grant to the party applying a certificate that the costs have not been paid. Upon production of this certificate to any justice, he may enforce the payment of such costs by warrant of distress; and in default of distress, may commit the person against whom the warrant was issued for any time not exceeding three calendar months, unless the costs, and all costs of the distress, and of the commitment, and conveying the party to prison, if so ordered (the amount being ascertained and stated in the commitment), shall be sooner paid.

After an appeal against any conviction has been decided, if it has been decided in favour of the respondents, the justices who convicted, or any other justice of the same county or borough, may issue such warrant of distress or commitment, for execution, as if no appeal had been brought.

The Court of Quarter Sessions has the power to state a "special case," upon a point of law, for the decision of the Queen's Bench, and when a special case is granted, the proceedings are removed by a writ of *certiorari*. Whenever, therefore, this writ is taken away by statute, the Sessions have no power to state a special case for the opinion of the Queen's Bench, unless with the express consent of both parties. As by the Weights and Measures Act, of 1878, the removal of the *certiorari* is repealed, the power to state a special case has thus been revived. The granting of a special case is entirely in the discretion of the Sessions, for the Court of Queen's

Bench will never compel them to grant one; but a case ought to be granted whenever the court are in doubt as to the legality of their decision. The Court of Quarter Sessions should decide the question one way or the other, but they may do so subject to the decision of the Queen's Bench upon the point reserved. The questions reserved must only be questions of law, for the decision of the Sessions upon questions of fact is final. The *certiorari* must be sued for within six calendar months after the order of the Sessions. If a case is not stated correctly, or in accordance with the rules laid down upon the subject, it may be sent back to the Sessions to be re-stated, and the Sessions may have fresh evidence if they choose to have it, and should hear it, if it is tendered. The effect of granting a special case is to make the judgment of the Queen's Bench that of the Sessions, who will be considered as adopting it.

There is also a power of reference to one of the superior courts of law by stating a special case without going to the Sessions previously, &c. This is done in accordance with the provisions of s. 11 of the 12 & 13 Vict. c. 45, which enacts, that "at any time after notice given of appeal to any Court of Quarter Sessions against any judgment or other matter for which the remedy is by appeal, it shall be lawful for the parties, by consent, and by order of any judge of one of the superior courts of law at Westminster, to state the facts of the case in the form of a special case for the opinion of such superior court, and to agree that a judgment in conformity with the decision of such court, and for such costs as such court shall adjudge, may be entered on motion, by either party, at the sessions next, or next but one, after such decision shall have been given; and such judgment shall and may be entered accordingly, and shall be of the same effect in all respects as if the same had been given by the Court of Quarter Sessions upon an appeal duly entered and continued." No recognizance entered into for the prosecution and trial of any appeal is to be deemed to be forfeited by such an agreement for stating a special case (12 & 13 Vict. c. 45, s. 16).

Further particulars with reference to the practice in cases of appeal, and legal decisions upon the points arising upon the stating of a special case, will be found in the books upon the jurisdiction and practice of the Court of Quarter Sessions.

(3.) *Actions against Inspectors and Others.*

Provision as to action against persons acting in execution of Weights and Measures Act, 1878. 41 & 42 Vict. c. 49, s. 61.

In an action for any act done in pursuance or execution, or intended execution, of the Weights and Measures Act, 1878, or in respect of any alleged neglect or default in the execution of that Act, tender of amends before the action is commenced may in lieu of or in addition to any other plea be pleaded, if the action was commenced after such tender, or is proceeded with after payment into court of any money in satisfaction of the plaintiff's claim. If the action is commenced after such tender, or is proceeded with after such payment, and the plaintiff does not recover more than the sum tendered or paid respectively, the plaintiff shall not recover any costs incurred after such tender or payment, and the defendant shall be entitled to his costs, to be taxed as between solicitor and client, as from the time of such tender or payment; but this provision shall not affect costs on any injunction in the action.

This section is an adaptation of s. 40 of 5 & 6 Will. IV. c. 63, to the modern changes in legal procedure. It gives only powers of tendering amends, and does not make *bona fides* a defence to the action (*Thomas* v. *Stephenson*, 22 L. J., Q. B., 258; 2 E. & B., 108; 17 Jur. 597; 1 W. R., 325; 17 J. P., 537).

CHAPTER X.

MISCELLANEOUS.

(1.) *Legal Provisions.*

Orders in Council.
41 & 42 Vict. c. 49, s. 63.

It shall be lawful for Her Majesty in Council from time to time to make orders for the purposes of the Weights and Measures Act, 1878, and to revoke and vary any such order.

All Orders in Council made under that Act shall be published in the London, Edinburgh, and Dublin Gazettes, and shall be forthwith laid before both Houses of Parliament, and shall have full effect as part of that Act.

Effect of schedules.
41 & 42 Vict. c. 49, s. 64.

The schedules to the Weights and Measures Act, 1878, with the notes thereto, shall be construed and have effect as part of that Act.

Construction of Acts referring to repealed enactments.
41 & 42 Vict. c. 49, s. 65.

Where an enactment refers to any Act repealed by the Weights and Measures Act, 1878, or to any enactment thereof, the same shall be construed to refer to the said Act or to the corresponding enactment of that Act.

Orders in Council.—There was a similar provision in the Standards of Weights, Measures, and Coinage Act, 1866 (29 & 30 Vict. c. 82, s. 8).

(2.) *Savings and Definitions.*

Saving as to models of gas holders under 22 & 23 Vict. c. 66;
41 & 42 Vict. c. 49, s. 66.

Nothing in the Weights and Measures Act, 1878, shall affect the validity of the models of gas holders verified and deposited in the standards department of the Board of Trade, in pursuance of the Act of the session of the twenty-second and twenty-third years of the reign of Her present Majesty, chapter sixty-six, intituled, "An Act for regulating measures used in sales of gas," and of the Acts amending the same; and the provisions of the Weights and Measures Act, 1878, with respect to Board of Trade standards, shall apply to such models; and the provisions of the Weights and Measures Act, 1878, with respect to defining the amount of error to be tolerated in local standards when verified or re-verified, shall apply to defining the amount of error to be tolerated in such copies of the said models of gas holders as are provided by any justices, council commissioners, or other local authority in pursuance of the said Acts.

Nothing in the Weights and Measures Act, 1878, shall extend to prohibit, defeat, injure, or lessen the rights granted by charter to the master, wardens, and commonalty of the mystery of Founders of the city of London. *Saving as to rights of the Founders Company. 41 & 42 Vict. c. 49, s. 67.*

Nothing in the Weights and Measures Act, 1878, shall prohibit, defeat, injure, or lessen the right of the mayor and commonalty and citizens of the city of London, or of the Lord Mayor of the city of London for the time being, with respect to the stamping or sealing of weights and measures, or with respect to the gauging of wine or oil, or other gaugeable liquors. *Saving as to London. 41 & 42 Vict. c. 49, s. 68.*

Nothing in the Weights and Measures Act, 1878, shall extend to supersede, limit, take away, lessen, or prevent the authority which any person or body politic or corporate, or any person appointed at any court leet for any hundred or manor, or any jury or ward inquest, may have or possess for the examining, regulating, seizing, breaking, or destroying any weights, balances, or measures within their respective jurisdictions, and for the purposes of this section the court of burgesses of the city of Westminster shall be deemed to be a body politic, and nothing in the said Act shall be deemed to repeal or *Act not to abridge the power of the leet jury, &c. 41 & 42 Vict. c. 49, s. 69.*

G

supersede the Acts relating to that court, or lessen, diminish, or alter the powers of the same.

Models of gas holders.—This section has been inserted because there were certain provisions of the Standards of Weights, Measures, and Coinage Act, 1866, now repealed, which applied to the gas standards under the Sale of Gas Act (22 & 23 Vict. c. 66).

The savings, as to the City of London, ward inquests, and the city of Westminster, were inserted in the Bill during its passage through Parliament. Ward inquests were believed to be abolished, and the general reservation of rights given in section 54 of the Act would have probably met the cases of London and Westminster. In the Act of 1835 there was a reservation of the rights of the Universities of Oxford and Cambridge to supervise weights and measures in the city of Oxford and the town of Cambridge, and power was given to the Chancellor or Vice-Chancellor of such University to appoint an inspector or inspectors of weights and measures with equal power and authority to those appointed by the justices (5 & 6 Will. IV. c. 63, s. 44). This saving is omitted in the Act of 1878. In the case of Cambridge, the University authorities have surrendered their power in respect to weights and measures; and as regards Oxford, where the University has surrendered its power for a limited time only, its rights will be sufficiently reserved by the general reservation in section 54 of the Act of 1878, which has already been considered.

The power of the leet jury.—The court leet is a court of record held once in the year, and not oftener, within a particular hundred, lordship, or manor. The earliest recorded mode of inspection of weights and measures was that under the authority of courts leet. The duties of inspection were discharged by annoyance juries appointed by these courts, who adjudicated upon all cases of false or unjust weights, measures, and balances. They also appointed separate district officers for the verification of weights and measures within their respective districts. The word "leet" is from the Saxon *Lathe, Leth. Loethe,* a sub-division of the county; and the court leet is the court of the lathe, just as the county court is the court of the county. The court leet has power to inquire into weights and measures within its jurisdiction; but the leet jury cannot in general enter shops, &c., to examine the weights, measures, and scales. The party must be proceeded against by summons (*Moore* v. *Wicker,* Andrews, 47). There may, however, be a custom in a manor to make such an entry, and then it will be lawful; and the custom may also extend for the jury to break and destroy weights and measures (*Willcock* v. *Windsor,* 3 B. & Ad. 43).

The county inspectors of weights and measures are not excluded from manors within their district, but have concurrent jurisdiction with the officials appointed by the courts leet.

The actual method in which this authority over inspection of weights and measures by courts leet was exercised by law, and the duties of inspection discharged by annoyance juries, may be seen in the case of the city of Westminster, as laid down in the Act passed in 1756 (29 Geo. II. c. 25), and in one passed two years later, in 1758 (31 Geo. II. c. 17). In 1861, however, the appointment of annoyance juries in the city of Westminster ceased, and provision was made for the appointment of inspectors of weights and measures by the dean and court of burgesses. The powers and rights with regard to weights and measures, now exercised by the court of burgesses within the City of Westminster, which are expressly reserved by section 69 of the Weights and Measures Act, 1878, will be found for the most part contained in the provisions of the statute 24 & 25 Vict. c. 78, which, for the benefit of persons residing in Westminster, is given in the *Appendix*.

In the Weights and Measures Act, 1878, unless the context otherwise requires,— *Definitions. 41 & 42 Vict. c. 49, s. 70.*

The expression "the Summary Jurisdiction Act," means the Act of the session of the eleventh and twelfth years of the reign of Her present Majesty, chapter forty-three, intituled "An Act to facilitate the performance of the duties of justices of the peace out of sessions within England and Wales with respect to summary convictions and orders," inclusive of any Acts amending the same.

The expression "court of summary jurisdiction" means any justice or justices of the peace, metropolitan police magistrate, stipendiary or other magistrate or officer, by whatever name called, to whom jurisdiction is given by the Summary Jurisdiction Act or any Acts therein referred to:

The expression "quarter sessions" includes general sessions:

The expression "Treasury" means the Commissioners of Her Majesty's Treasury:

The expression "person" includes a body corporate:

The expression "stamping" includes casting, engraving, etching, branding, or otherwise marking, in such manner as to be so far as practicable indelible, and the expression "stamp," and other expressions relating thereto, shall be construed accordingly:

The expression "coin weight" means a weight used or intended to be used for weighing coin:

The expression "Weights and Measures Act, 1835," means the Act of the fifth and sixth years of the reign of King William the Fourth, chapter sixty-three, intituled "An Act to repeal an Act of the fourth and fifth years of His present Majesty relating to weights and measures, and to make other provisions instead thereof."

These definitions have nearly all been inserted in the notes appended to the sections in which they occur.

(3.) *Repeal.*

Repeal.
41 & 42 Vict.
c. 49, s. 86.

The Acts mentioned in the first part of the Sixth Schedule to the Weights and Measures Act, 1878, are hereby repealed to the extent in the third column of that schedule mentioned; subject to the following qualification, that is to say, that so much of the said Acts as is set forth in the second part of that schedule shall be re-enacted in manner therein appearing, and shall be in force as if enacted in the body of the Weights and Measures Act, 1878.

Provided that,—

1. Every inspector appointed in pursuance of any enactment hereby repealed shall continue in office as if he had been appointed in pursuance of the Weights and Measures Act, 1878; and,
2. Any person holding office as examiner of weights and measures under any enactment repealed by the Weights and Measures Act, 1878, and not being an inspector of weights and measures within the meaning of that Act, shall continue in office and receive the same remuneration, and have the same powers and duties, and be subject to the same liabilities

and to the same power of dismissal as if that Act had not passed.

3. Every notice published in a Gazette in relation to coin weights, in pursuance of any enactment hereby repealed, shall continue in force.

4. All weights and measures duly verified and stamped in pursuance of any enactment hereby repealed, shall continue and be as valid as if they had been verified and stamped in pursuance of the Weights and Measures Act, 1878, and that although such weights or measures could not have been verified and stamped in pursuance of that Act; and all weights and measures which at the commencement of that Act may lawfully be used without being stamped with a stamp of verification or a stamp of their denomination, and which are required by that Act to be stamped with such a stamp, may, notwithstanding they are not so stamped, be used until the expiration of six months after the commencement of the said Act, without being subject to be seized or forfeited, and without rendering the person using or having possession of the same subject to any fine.

5. This repeal shall not affect—

(a.) The past operation of any enactment hereby repealed, nor anything duly done or suffered under any enactment hereby repealed; nor

(b.) Any right, privilege, obligation, or liability acquired, accrued, or incurred under any enactment hereby repealed; nor

(c.) Any penalty, forfeiture, or punishment incurred in respect of any offence committed against any enactment hereby repealed; nor

(d.) Any investigation, legal proceeding, or remedy in respect of any such right, privilege, obligation, liability, penalty, forfeiture, or punishment as aforesaid; and any such investigation, legal proceeding, and remedy may be carried on as if the Weights and Measures Act, 1878, had not passed; and

6. This repeal shall not revive any enactment, right, office, privilege, matter, or thing not in force or existing at the commencement of the Weights and Measures Act, 1878.

The sixth schedule to the Weights and Measures Act, 1878, is given in full in the *Appendix*. It will be noticed that the old Acts of 1795 and 1797 for weights, and of 1815 for measures, are included in this schedule, and are thus expressly repealed. But most of the provisions of these enactments were virtually repealed by the Act of 1835, which now in its turn gives place to the new Act of 1878. See *Stevenson* v. *Sheckle*, 12 J. P. 772, and 13 J. P. 86. The provisos which follow the repealing clause have nearly all been considered and inserted in the notes which have already been given. Inspectors already appointed are to continue in their office, and examiners of weights and measures under the old Acts, if any exist, may continue to carry on their duties. The Gazette notices in relation to coin weights, which will be found in the Appendix to the Fifth Annual Report of the Warden of the Standards, 1870–71, are to continue in force; and six months' grace is allowed for weights and measures, hitherto not requiring a stamp, to be stamped with the stamp of verification under the new Act.

(4.) *Miscellaneous Provisions*.

Continuance of inquisition recorded for ascertaining rents and tolls payable.
41 & 42 Vict. c. 49, s. 62.

Every inquisition which, in pursuance of any Act repealed by the Weights and Measures Act, 1878, has been taken for ascertaining the amount of contracts to be performed or rents to be paid in grain or malt, or in any other commodity or thing, or with reference to the measure or weight of any grain, malt, or other commodity or thing, and the amount of any toll, rate, or duty payable according to any weight or measure in use before the passing of the said Act, and has been enrolled of record in Her Majesty's Court of Exchequer, shall continue in force, and may be given in evidence in any legal proceeding, and the amount ascertained by

Miscellaneous.

such inquisition shall, when converted into imperial weights and measures, continue to be the rule of payment in regard to all such contracts, rents, tolls, rates or duties.

<small>This section provides for the continuance in force of inquisitions, taken in pursuance of sections 14 and 15 of the Weights and Measures Act, 1835 (5 & 6 Will. IV., c. 63, ss. 14, 15).</small>

The owners or managers of any public market in Great Britain where goods are exposed or kept for sale shall provide proper scales and balances, and weights and measures or other machines, for the purpose of weighing or measuring all goods sold, offered, or exposed for sale in any such market, and shall deposit the same at the office of the clerk or toll collector of such market, or some other convenient place, and shall have the accuracy of all such scales and balances, and weights and measures or other machines, tested at least twice in every year by the inspector of weights and measures of and for the county, borough, or place where the market is situate; Owners of markets to provide scales, &c. 41 & 42 Vict. c. 49, schedule.

All expenses attending the purchase, adjusting, and testing thereof shall be paid out of the moneys collected for tolls in the market;

Such clerk or toll collector shall at all reasonable times, whenever called upon so to do, weigh or measure all goods which have been sold, offered, or exposed for sale in any such market, upon payment of such reasonable sum as may from time to time be decided upon by the said owners or managers, subject to the approval and revision of the justices in general or quarter sessions assembled if such market be in England, or of the sheriff if it be in Scotland;

For every contravention of this section the offender shall be liable, on summary conviction, to a fine not exceeding five pounds.

Power to clerks of markets to inspect goods sold, &c.
41 & 42 Vict. c. 49, schedule.

Every clerk or toll collector of any public market in Great Britain, at all reasonable times, may weigh or measure all goods sold, offered, or exposed for sale in any such market; and if upon such weighing or measuring any such goods are found deficient in weight or measure, or otherwise contrary to the provisions of the Weights and Measures Act, 1878, such clerk or toll collector shall take the necessary proceedings for recovering any fine, to which the person selling, offering, or exposing for sale, or causing to be sold, offered, or exposed for sale. such goods, is liable, and the court convicting the offender may award out of the fine to such clerk or toll collector such reasonable remuneration as to the court seems fit.

For every offence against or disobedience to this section, the offender shall be liable, on summary conviction, to a fine not exceeding five pounds.

The above sections are in the second part of the sixth schedule to the Act of 1878, and they are re-enactments of similar provisions in the Weights and Measures Act of 1859 (22 & 23 Vict. c. 56. ss. 6, 7, 8, 12), which are now repealed. They relate incidentally only to weights and measures, and have been placed in a schedule so as to allow, when they have to be consolidated with the Acts to which they properly belong, of their being struck out of the schedule without altering a material part of the Act itself. These sections do not apply to Ireland.

Excise penalty for trader using false weights and scales in weighing his stock.
10 Geo. III., c. 44, s. 1.

After reciting that several traders subject to the survey of the officers of excise, who are required to keep just and sufficient scales and weights at the places where they carry on their respective trades, to be used in taking the account of their stocks, have used false, unjust, and insufficient scales and weights, to the great diminution of the revenue, and to the discouragement of the fair trader, it is enacted, that, if at any time after the twenty-fourth day of June, 1770, any trader, subject to the survey of any officer of excise, and required by the laws concerning the duties under the management of the commissioners of excise to keep sufficient and just scales and weights, shall, in the weighing his, her, or their stock or stocks, make use of, or cause, or permit or suffer to be used, any false, unjust, or insufficient scales or weights, with the intent to defraud His Majesty of the duties by the said laws respectively granted, that then, and in every such case, the party or parties offending shall forfeit the sum of one hundred pounds for every such offence.

If at any time or times after the first day of August, 1786, any trader or traders, subject to the survey of any officer or officers of the excise, or inland duties, and who is or are required by any law or laws relating to the duties of excise, or other duties under the management of the commissioners of excise, to keep just scales and weights, shall, before or after, or in the weighing of his, her, or their stock, or any part thereof, put or suffer, or cause or procure to be put, any other substance into the commodity or stock so to be weighed, whereby such officer or officers might be hindered or prevented from taking a just and true account of such stock, as is directed and prescribed by the several Acts of Parliament in that case made and provided, or shall forcibly obstruct or hinder, or shall by any act, service, or contrivance prevent or impede such officer, or procure or suffer him to be prevented or impeded in taking such just and true account of such stock or commodities as aforesaid, the party offending therein shall, for every such offence, forfeit and lose the sum of one hundred pounds. *Excise penalty for trader deceiving officers of excise. 26 Geo. III., c. 77, s. 8.*

If any trader subject to the survey of any officer of excise, and required by the laws concerning the duties under the management of the commissioners of excise to keep sufficient and just scales and weights, shall, in the weighing his, her, or their stock or stocks, make use of, or cause, or procure, or suffer to *Seizure by excise officer of unjust scales or weights used in weighing stock, and forfeiture.*

28 Geo. III., c. 37, s. 15.

be used, any false, unjust, or insufficient scales or weights, to the intent to defraud His Majesty of the duties by the said laws respectively granted, such scales and weights respectively shall be forfeited, and shall and may be seized by any officer or officers of excise.

A balance and weights to be kept in every corn mill.
36 Geo. III., c. 85, s. 1.

Every miller or other person keeping a mill for the grinding of corn, shall have in such mill a true and equal balance, with proper weights, according to the standard of the Exchequer; and every miller or other person as aforesaid, in whose mill shall be found no balance or weights, shall forfeit and pay a sum not exceeding twenty shillings.

The power of examining the balances and weights kept in a mill in accordance with the provisions of the above section was originally given to the examiners of weights and measures appointed under the Act of 1795, and if any of these officers exist their powers will still continue, their rights having been expressly reserved by the Act of 1878. If there are no such officers, these weights and balances will be subject to the examination of the inspectors of weights and measures as they will be upon premises used for trade within the meaning of section 19 of the Act of 1878 (41 & 42 Vict. c. 49, s. 19). This being so, the greater portion of the first section of the Act of 1796 has been repealed by the Weights and Measures Act, 1878, being no longer required. The unrepealed part of the section, providing that a balance and weights are to be kept in every corn mill, is given above.

Proper weights and measures to be provided at markets and fairs.
10 & 11 Vict. c. 14, s. 21.

The undertakers, or persons authorised by a special Act to construct or regulate a market or fair, shall provide sufficient and proper weighing-houses, or places for weighing or measuring the commodities sold in the market or fair, and shall keep therein proper weights, scales, and measures according to the standard weights and measures for the time being, for weighing such commodities as aforesaid, and shall appoint proper persons to attend to the weighing or measuring such commodities at all times during which the market or fair is holden.

Machines for

They shall also provide sufficient and proper buildings or places for weighing carts in which goods are brought for sale within the market or fair or the prescribed limits, and shall keep therein machines and weights proper for that purpose, and shall from time to time appoint a person in every such building or place to afford the use of such machines to the public by weighing such carts, with or without their loading, as may be required.

<small>Weighing carts laden with goods also to be provided. 10 & 11 Vict. c. 14, s. 24.</small>

These two sections are taken from the Markets and Fairs Clauses Act, 1847. There are various other provisions in the Act with regard to the weighing of goods and carts at markets and fairs. Every person selling in a market or fair is, if required so to do by the buyer, to weigh or measure the article sold by the public weights, measures, or scales, provided for that purpose; and the driver of every cart in which goods are brought for sale within the market or fair is, at the request of the buyer or seller of such goods, to take the cart to be weighed at one of the public weighing machines either with or without its load, or both. Ample provisions are also made for the punishment of frauds committed in the weighing, either by the drivers of carts, by buyers or sellers, or by the persons in charge of the weighing machine. See 10 & 11 Vict. c. 14.

The Weights and Measures Act, or any Act for the time being in force relating to weights and measures, shall apply to the weights used in any mine to which the "Coal Mines Regulation Act, 1872," applies, for determining the wages payable to any person employed in such mine, according to the weight of the mineral gotten by such person, in like manner as it applies to weights used for the sale of any article, and the inspector of weights and measures for the district appointed under the Weights and Measures Act, shall accordingly from time to time, but without unnecessarily impeding or interrupting the working of the mine, inspect and examine, in manner directed by the Weights and Measures Act, the weighing machines and weights used for mines to which the "Coal Mines Regulation Act, 1872," applies, or the measures or gauges used for such mines: provided that nothing in this section shall

<small>Application of Weights and Measures Act to weights used in mines, &c. 35 & 36 Vict. c. 76, s. 19.</small>

prevent the use of the measures and gauges ordinarily used in such mine.

The Weights and Measures Act here referred to was the Weights and Measures Act, 1835; but as the section expressly applies to "any Act for the time being in force relating to weights and measures," the Act of 1878 will now take its place. The various provisions with regard to the payment of persons employed in mines according to the weight of the mineral gotten by them, and for the correct and equitable weighing of such mineral, will be found in the other sections of the Coal Mines Regulation Act, 1872 (35 & 36 Vict. c. 76).

CHAPTER XI.

SCOTLAND AND IRELAND.

(1.) *Scotland.*

THE Weights and Measures Act, 1878, shall apply to Scotland with the following modifications:

In the application of the Act to Scotland, the expression "rents and tolls" includes all stipends, feu duties, customs, casualties, and other demands whatsoever payable in grain, malt, or meal, or any other commodity or thing.

The fiars prices of all grain in every county shall be struck by the imperial quarter, and all other returns of the prices of grain shall be set forth by the same, without reference to any other measure whatsoever.

Any person who acts in contravention of this provision shall be liable to a fine not exceeding five pounds.

Application of imperial weights and measures to tolls, &c. 41 & 42 Vict. c. 49, s. 71.

Shall apply to Scotland.—There were several attempts before the Union to secure by legislation the uniformity of Weights and Measures in Scotland, but without much success. By the Act of Union it was provided that "the same weights and measures should be used throughout the United Kingdom as were then established in England." In 1824, to secure the advantages of uniformity the imperial standards were introduced, and were established for the whole of the United Kingdom of Great Britain and Ireland (5 Geo. IV. c. 74). In 1825 this Act was amended, and again in 1835 the Weights and Measures Act of that year included the whole of the United Kingdom. The sections applying to Scotland in the present Act are for the most part re-enactments of similar provisions in the Act of 1835, with such modifications as the change in the general law rendered necessary.

The expression "rents and tolls."—As a matter of fact no such phrase as "rents and tolls" is to be found in the body of the Act. But the reference may either be to section 19, where the words "tolls" and "duties" are found, or to section 62, which deals with "contracts, rents, tolls, rates or duties." By disregarding the quotation marks and taking it as "rents" or "tolls," the definition applies to both sections, and was probably inserted as a ready method of adapting these clauses to Scotland.

Fiars prices.—This is an exact re-enactment of 5 & 6 Will. IV. c. 63, s. 16. The fiars prices are the prices of grain in the different counties fixed by the sheriffs respectively, with the aid of juries, for the purpose of establishing a standard by which, for the year, contracts may be governed.

Recovery and application of penalties.
41 & 42 Vict. c. 49, s. 72.

All offences under the Weights and Measures Act, 1878, which may be prosecuted, and all fines and forfeitures under that Act which may be recovered on summary conviction, may in Scotland be prosecuted or recovered, with expenses, before the sheriff or sheriff substitute or two or more justices of the peace of the county, or the magistrates of the burgh wherein the offence was committed or the offender resides, at the instance either of the procurator fiscal or of any person who prosecutes.

Every person found liable in Scotland in any fine recoverable summarily under the said Act shall, failing payment thereof immediate or within a specified time, as the case may be, and expenses, be liable to be imprisoned for a term not exceeding sixty days, and the conviction and warrant may be in the form number three of Schedule K of the Summary Procedure Act, 1864.

All fines and forfeitures so recovered, subject to any payment made to the informer, shall be paid as follows:

(*a.*) To the Queen's and Lord Treasurer's Remembrancer, on behalf of Her Majesty, when the court is the sheriff court:

(*b.*) To the collector of county rates, in aid of the county general assessment, when the court is the justice of the peace court:

(*c.*) To the treasurer of the burgh, in aid of the funds of the burgh, when the court is a burgh court:

(*d.*) To the treasurer of the board of police, or commissioners of police, in aid of the police funds, when the court is a police court.

This section adapts the provisions of section 37 of the Act of 1835 to the modern procedure.

Schedule K of the Summary Procedure Act, 1864.—All offences under the Weights and Measures Act may be prosecuted, and all fines and forfeitures under that Act may be recovered on summary conviction before a court of summary jurisdiction in manner provided by the Summary Jurisdiction Act. For the purposes of Scotland the Summary Jurisdiction Act means the Summary Procedure Act, 1864, which is an Act providing for uniformity of process in summary criminal prosecutions and prosecutions for penalties in the inferior courts in Scotland (27 & 28 Vict. c. 53). See *post*.

An appeal against a conviction under the Weights and Measures Act, 1878, in Scotland shall be to the Court of Justiciary at the next circuit court, or where there are no circuit courts, to the High Court of Justiciary at Edinburgh, and not otherwise, and such appeal may be made in the manner and under the rules, limitations, and conditions contained in the Act of the twentieth year of the reign of King George the Second, chapter forty-three, intituled, "An Act for taking away and abolishing heritable jurisdictions in Scotland," or as near thereto as circumstances admit; with this variation, that the appellant shall find caution to pay the fine and expenses awarded against him by the conviction or order appealed from, together with any additional expenses awarded by the court dismissing the appeal.

Appeal.
41 & 42 Vict. c. 49, s. 73.

Where there are no circuit courts.—The Court of Justiciary is the supreme criminal court of Scotland. It holds circuit courts at different towns twice a year, with an additional circuit court at Glasgow, for the trial of cases from all counties in Scotland except the three Lothians, Peebles, Orkney and Shetland. Cases from the latter are tried at the High Court in Edinburgh. See Macdonald's "Criminal Law of Scotland."

And not otherwise.—It may be a question whether these words do or do not exclude the very useful and important provisions of the "Summary Prosecutions Appeals (Scotland) Act, 1875," as to the statement of a "case" by the inferior judge upon questions of law for the opinion of the High Court of Justiciary at Edinburgh. See 38 & 39 Vict. c. 62.

Act of the twentieth year of the reign of King George the Second, chapter forty-three.—This Act, with the view of rendering the jurisdiction of the circuit courts in Scotland more useful, enacted, that it should be lawful for any party conceiving himself aggrieved by decree, sentence, or judgment of the sheriff's or stewart's court of any county, or of any royal burgh (and of certain other inferior courts), in certain criminal and civil cases, to seek relief against the same by appeal to the next circuit court of the circuit wherein such county, stewarty, or royal burgh should lie, but so as no such appeal should be competent before a final decree, sentence, or judgment pronounced; that it should be lawful to take and enter such appeal in open court at the time of pronouncing such decree, sentence, or judgment, or at any time thereafter within *ten* days, by lodging the same in the hands of the clerk of the court, and serving the adverse party with a duplicate thereof personally or at his dwelling-house, or his procurator or agent in the case, and serving in like manner the inferior judge himself in case the appeal should contain any conclusion against him by way of censure or reparation of damages for alleged wilful injustice, oppression, or malversation; that such service should be sufficient summons to oblige the respondents to attend and answer at the next circuit court which should happen to be held *fifteen* days at least after such service; that, thereupon, the judge or judges at such circuit court should and might proceed to hear and determine any such appeal by the like rules of law and justice as the Court of Session or Court of Justiciary respectively might then determine in suspensions of decrees, sentences, or judgments of such inferior courts; but the said circuit court should proceed therein in a summary way; that in case they should find the reasons of any such appeal not to be relevant or not instructed, or should determine against the party so appealing, the said judge or judges should condemn the appellant in such costs as the court should think proper to be paid to the other party, not exceeding the real costs *bonâ fide* expended by such party; and that the decree, sentence, or judgment of such circuit court in any of the cases aforesaid should be final; that wherever such appeal should be brought, the complainer at the same time he entered his appeal should lodge in the hands of the clerk of court from which the appeal was taken a bond, with a sufficient cautioner for answering and abiding by the judgment of the circuit court, and for paying the costs, if any should be by that court awarded (*and in paying the fine and expenses awarded against him by the conviction or order appealed from, see section above*); and that the clerk of court should be answerable for the sufficiency of such cautioner; and that in case such circuit court should, in proceeding upon such appeal, find any such difficulty to arise that by means thereof such circuit court could not proceed to the determination of the same consistently with justice and the nature of the case, it should be lawful for such circuit court to certify such appeal, together with the reasons of such difficulty and the proceedings thereupon had before such circuit court to the Court of Session or Court of Justiciary respectively, which courts were thereby respectively authorised and required to proceed in and determine the same.

In the application of the Weights and Measures Act, 1878, to Scotland,— *Definitions. 41 & 42 Vict. c. 49, s. 74.*

The expression "enter into a recognizance" means grant a bond of caution:

The expression "any court of record" includes the Court of Session and the ordinary sheriff court:

The expression "burgh" shall include royal burgh and parliamentary burgh:

The expression "plaintiff" means pursuer, and the expression "defendant" means defender:

The expression "solicitor" means writer or agent:

The expression "Summary Jurisdiction Act" means the Summary Procedure Act, 1864, inclusive of any Act amending the same.

These definitions are required to adapt the several provisions of the Act to the legal phraseology used in Scotland.

Burgh.—This definition, as will be seen, is made to include a "Parliamentary burgh," which has not hitherto been recognised in the Weights and Measures Acts as a separate area for the appointment of inspectors.

The expression "Summary Jurisdiction Act" means, &c.—The effect of this clause, in conjunction with sects. 56 and 72 *ante*, is, that the offences, fines and forfeitures under the Act may, in Scotland, be prosecuted and recovered in a summary way before the sheriff, or sheriff substitute, or two or more justices of the peace of the county, or the magistrates of the burgh wherein the offence was committed, or the offender resides, in manner provided by "the Summary Procedure Act, 1864," inclusive of any Act amending the same. Bearing upon the subject of procedure, this Act of 1864, *inter alia*, provides in effect as follows:

Complaint.—This may be in one or other of the short forms annexed to the Act; it being sufficient to refer in the complaint to the particular Act and section declaring the offence and imposing the penalty, without setting out the words at length ; sect. 4.

Objections to complaint.—Such are not to be allowed for alleged defect in substance or form, or for a variance between the complaint and the evidence adduced, *not changing the character of the offence charged*: and power is given to the court to adjourn the hearing and to direct amendment upon the complaint, *not changing the character of the offence*, in cases where any objection or variance appears to have deceived or misled the respondent; sect. 5.

H

Citation of respondent.—The court, on the complaint being laid before it, may grant warrant to cite by delivering to respondent personally a copy of the complaint with the court's warrant of citation, or where he cannot be found, by leaving such at his usual place of abode, to appear before the court on induciæ of not less than forty-eight hours; annexing to such warrant or citation a warrant to cite witnesses and havers for both parties; sect. 6.

Failure of respondent's appearance.—If the respondent fails to appear after being cited the court may, upon proof of due citation, issue its warrant in the second instance for his apprehension and interim detention; or may adjourn the hearing to a future diet with liberty to the respondent to appear thereat; and may appoint intimation of the adjourned diet to be made to him: and where the complaint concludes for a pecuniary penalty only in the first instance, the court may, without the presence of the respondent, hear and dispose of the complaint (sect. 7).

Adjournment of hearing on request of apprehended respondent.—A respondent, brought before the court by a warrant of apprehension, shall be entitled to require a copy of the complaint and an adjournment of the hearing for a period of not less than forty-eight hours, if the requisition be made before the examination of any witness on the merits has commenced, and no copy of the complaint has been delivered to him personally forty-eight hours before the hearing (sect. 11).

Adjournment of hearing and detention of respondent.—Subject to the last provision no adjournment of the hearing shall take place when the respondent pleads not guilty, or at any other stage of the proceedings, unless the court think fit to order an adjournment: provided that, where the respondent is brought up on warrant of apprehension, the court may grant warrant to detain him in prison until the time of adjourned hearing, or till he finds caution to appear at future diets of court (sect. 12).

Hearing, &c.—When the respondent is *present* at the hearing, the substance of the complaint shall be read to him, and he shall be required to plead, and he may then state objections to the competency or relevancy of the complaint or proceedings; if no objections are stated, or, if stated, repelled or obviated by amendment of the complaint, or adjournment of the diet, the respondent's plea shall then, or at such adjourned diet, be recorded; and, if the plea be guilty, shall be signed by him, or, if he cannot write, by a judge or clerk of court; if the plea be not guilty, the prosecutor shall proceed to establish his case by competent evidence; and the respondent may lead such evidence as is competent; after which, the court shall pronounce judgment at the same, or any adjourned diet. When the respondent is *absent* the court shall not pronounce judgment against him until the complaint is established by competent evidence. It shall not be necessary, in any proceeding under the Act, to preserve or record a note of the evidence adduced; but the record must set forth the respondent's plea, if any, the names of the witnesses, if any, examined on oath or affirmation, with a note of any documentary evidence that may be put in: and the forms of proceedings prescribed by the Act may be, in whole or in part, written or printed. In cases in which, under any Act, any matter which may be dealt with under this Act is cognizable by two or more justices, it shall be sufficient that any warrant or proceeding before or after the conviction or judgment shall be subscribed by one justice; but the conviction or judgment shall, in all cases, be signed by such number of justices present at the hearing and concurring in the result thereof as may be required by such Act; and, in case of an equal division of opinion among the justices present, the complaint shall be held to be not proved, and judgment given for the respondent; and all warrants of citation under the Act may be signed by the clerk of court without being laid before a judge. In all cases of complaint for recovery of any penalty the court may award expenses without the same being prayed for in such complaint; but expenses shall not be awarded to or against any public prosecutor, or party prosecuting under any Act for the public interest, unless such expenses are authorised by such Act; provided that, when the complaint is at the instance of a private complainer, the court may award expenses to the successful party; and these may be recovered and imprisonment awarded in default of payment, or recovery, in the form mentioned in the Act, or by a separate judgment for expenses in an appropriate form. In all cases in which no time is limited for instituting complaints for recovery of any penalty, or for conviction for any statutory offence punishable on summary conviction in the Act relating to each particular case, such complaint shall be instituted within six months from the time when the matter of such complaint arose (sects. 14, 15, 16, 17, 21, 22, 24).

A sheriff or sheriff substitute shall have the same power in relation to a local comparison of standards, and to the inspection, comparison, seizure, and detention of weights and measures, and to entry for that purpose, as is given by the Weights and Measures Act, 1878, to a justice of the peace. *Power of sheriff. 41 & 42 Vict. c. 49, s. 75*

For the local comparison of standards before a justice, see section 41 of the Act of 1878;—Administration, *ante*; and for the inspection of weights and measures, see section 48 of the same Act;—Inspectors and their duties, *ante*. The sheriff in Scotland has also to approve and revise the sum to be paid to a clerk or toll collector of a market for weighing or measuring the goods which have been sold or offered for sale in the market, and which are brought to the said clerk or toll collector to be weighed or measured. See Miscellaneous, *ante*.

The "local authority" in Scotland is declared by the fourth schedule of the Act of 1878, which will be found in the *Appendix*, to be, in the county the "justices in general or quarter sessions assembled," and in the burgh the "magistrates." The "local rate" in the same schedule is declared to be, in the county the "county general assessment," and in the burgh. the " police assessment." This provision as to the local rate is in accordance with the existing pract ce. The Act of 1835 directed the expenses to be raised together with the land-tax (5 & 6 Will. IV. c. 63. s. 22).

It may be useful to note the following cases which have been decided in the courts of law in Scotland upon the former Weights and Measures Acts.

A landlord and tenant entered into missives of lease, in which the rent was fixed at a half boll of wheat, three firlots of barley, and six pecks of oats for each Scotch acre. payable by the fiars prices; but the proportions were not expressed which these measures bore to the imperial standard measure; and the landlord, under whose direction the missive of lease had been framed, raised an action, after the tenant had entered into possession, to reduce the lease, libelling upon the Act 5 Geo. IV. c. 74, ordaining uniformity of weights and measures. It was held that the statute did not apply to contracts of lease (*Henry* v. *McEwan*, 25 May, 1832, 10 S. D. & B. 572; affirmed 9 Aug. 1834, 7 W. & S. App. Cases, 411).

In a lease the rent was made partly dependent upon the price of cheese " per stone, according to the highest fiars of the county." It was held that this expression was capable of construction, and that, in the circumstances, it meant the market price per *tron* stone —a local measure of 24 lbs.—and that the tenant was not entitled to insist either upon applying the price of the imperial stone of 14 lbs., or on having the contract annulled under the Weights and Measures Acts (*Miller* v. *Muir*. Feb. 3, 1860. C. S. C. 2nd Ser. 22nd Vol. 660).

The sale of an article by a vessel other than an imperial measure is legal, provided that such vessel is not represented as containing any amount of imperial measure. The representation that a vessel contains a particular amount of imperial measure, is to be inferred from the practice of selling by such vessel, when it contains such amount of imperial measure (*Davie* v. *Robertson*. July, 19, 1847. Arkley, 336).

A sale of goods at a price " per stone, and each stone to consist of 22 lbs.," is not null under the Weights and Measures Acts, the stone stipulated being stated to be a multiple of a legal weight (*Robertson* v. *Good*. June 25, 1858. C. S. C. (2nd Ser), 20th Vol. 1170).

A sentence under the Weights and Measures Act, 1835, was held to be reviewable on objections to legality of the procedure; and if the inspector was disqualified, the conviction was bad (*Robertson* v. *Hart*. Dec. 24, 1842. 1 Broun's Justiciary Rep. 468).

In an appeal before the circuit court against a conviction obtained upon a complaint charging a contravention of section 21 of the Weights and Measures Act, 1835, held that the use of weights other than those authorised by the Weights and Measures Acts for the purpose of weighing ironstone wrought by miners, in order to determine their wages, amounted to a contravention of the above section, and objection to the relevancy of the complaint charging such a use as amounting to said contravention repelled, and the appeal dismissed (*Paterson* v. *Robertson*, 2 Couper's Justiciary Reports, 131).

(2) *Ireland.*

The Weights and Measures Act, 1878, shall apply to Ireland with the following modifications:

In Ireland, every contract, bargain, sale, or dealing— *Contracts to be made by denominations of imperial weight, otherwise to be void. 41 & 42 Vict. c. 49, s. 76.*

For any quantity of corn, grain, pulses, potatoes, hay, straw, flax, roots, carcases of beef or mutton, butter, wool, or dead pigs, sold, delivered, or agreed for;

Or for any quantity of any other commodity sold, delivered, or agreed for by weight (not being a commodity which may by law be sold by the troy ounce or by apothecaries weight),

shall be made or had by one of the following denominations of imperial weight; namely,

 the ounce avoirdupois;
 the imperial pound of sixteen ounces;
 the stone of fourteen pounds;
 the quarter hundred of twenty-eight pounds;
 the half hundred of fifty-six pounds;
 the hundredweight of one hundred and twelve pounds; or
 the ton of twenty hundredweight;

and not by any local or customary denomination of weight whatsoever, otherwise such contract bargain, sale, or dealing, shall be void:

Provided always, that nothing in the present section shall be deemed to prevent the use in any contract, bargain, sale, or dealing of the denomination of the quarter, half, or other aliquot part of the ounce, pound, or other denomination aforesaid, or shall be deemed to extend to any contract, bargain, sale, or dealing, relating to standing or growing crops.

Shall apply to Ireland.—Hitherto the law relating to weights and measures in Ireland has been contained in two Acts, the " Weights and Measures (Ireland) Act, 1860," and the " Weights and Measures (Ireland) Amendment Act, 1862." These two Acts have now been for the most part repealed, and whilst some of their provisions have been made to apply to the whole of the United Kingdom, others will be found in the following sections which are modifications of the general law, suitable to the particular case of Ireland. The portion of the Weights and Measures (Ireland) Amendment Act, 1862, which has not been repealed, will be found at the end of this chapter. The section given above is a reproduction of a similar section in the Act of 1862 (25 & 26 Vict. c. 76, s. 12).

Mode of weighing.
41 & 42 Vict. c. 49, s. 77.

In Ireland every article sold by weight shall, if weighed, be weighed in full net standing beam ; and for the purposes of every contract, bargain, sale, or dealing, the weight so ascertained shall be deemed the true weight of the article, and no deduction or allowance for tret or beamage, or on any other account, or under any other name whatsoever, the weight of any sack, vessel, or other covering in which such article may be contained alone excepted, shall be claimed or made by any purchaser on any pretext whatever, under a penalty not exceeding five pounds.

A proceeding for the recovery of a penalty under this section shall be begun within three months after the offence is committed.

This section is a re-enactment of a similar clause in the Act of 1862 (25 & 26 Vict. c. 76, ss. 13–18).

The deductions prohibited in this section are deductions from the *weight*, not the *price*, of any article sold by weight (*Megarry* v. *McCullagh*, 14 Ir. C. L. R. 151).

Providing of local standards and sub-standards.
41 & 42 Vict. c. 49, s. 78.

1. The local authority in Ireland shall provide one complete set of local standards for their county or borough; also so many copies in iron or other sufficient material of the local standards.
2. The said copies of the local standards when duly verified as hereinafter mentioned, shall be the local sub-standards, and shall be used

for the verification of weights and measures brought by the public for verification as if they were local standards.
3. Not less than one set of local sub-standards, and one set of accurate scales, shall be provided for each petty sessions district in a county, and not less than two such sets shall be provided for a borough.
4. The local authority shall have the local standards from time to time duly compared and re-verified in manner directed by the Weights and Measures Act, 1878.
5. The Commissioners of the Dublin Metropolitan Police shall not be under any obligation to provide local standards, but they may, with the assent of the chief secretary or under secretary to the Lord Lieutenant, procure such sub-standards, scales, and stamps, as they think necessary for the purposes of the said Act in the district for which they are the local authority.

Local authority.—The definition of the local authority will be found in the fourth schedule of the Act given in the *Appendix*. In the county it is the "grand jury acting at any assizes or presenting term," the local rate being the "presentments to be made by the grand jury." The expression "county" is to include a "riding and a county of a city and a county of a town." The county of Dublin is not to include any portion of the police district of Dublin metropolis; and the two constabulary districts of the county of Galway are to be counties for the purposes of the Weights and Measures Act. The local authority for such portion of the police district of Dublin metropolis as is without the municipal boundary of the borough of Dublin is the "Commissioners of the Dublin Metropolitan Police," the local rate being the "funds applicable to defray the expenses" of the said police. In boroughs the local authority is the "Town Council," and the local rate is the "rate to be levied by the council, or if the borough is liable to county cess and no rate is levied on the borough, the county cess of the county in which the borough or the larger part thereof is situate." The expression "borough" means any borough or town corporate; and in the borough of Dublin the rate to be levied by the Council means the improvement rate.

Inquiry by judge of assize and chairman of quarter sessions as to provision of local standards and sub-standards. 41 & 42 Vict. c. 49, s. 79.

In Ireland, in every year—

a. in the case of a county, the judge of assize at the first assizes held for the county by inquiry of the foreman of the grand jury; and

b. in the case of every borough in a county, the recorder of the borough, or, if there be no recorder, the chairman of the quarter sessions for that county, at the quarter sessions held next after the twenty-fifth day of March,

shall inquire whether one complete set of local standards, and a sufficient number of local sub-standards of weights and measures, and a sufficient number of scales and stamps (for verification), have been provided in such county or in such borough.

If it appear to the judge or chairman upon such inquiry, that the same have not been so provided, he shall forthwith order the proper officer to provide a complete set of local standards, and such sub-standards, scales, and stamps, as appear to the judge or chairman making the order to be sufficient for the purposes of the Weights and Measures Act, 1878, and that order shall have the effect in the case of a county, of a presentment on the county for, and in the case of a borough, of an order on the council of the borough to raise by way of rate, the sum necessary to execute the order, and the said officer shall within three months after he receives the order fully execute the same, and in default shall be liable to a fine not exceeding twenty pounds.

The proper officer shall, in the case of a county, be the treasurer of the county, and in the case of a borough, the town clerk or other proper officer of the borough.

This section combines the provisions of clauses 9 and 10 in the Act of 1862 (25 & 26 Vict. c. 76, ss. 9, 10).

Expenses incurred by any member of the Royal Irish Constabulary, as an ex-officio inspector of weights and measures, in the execution of the Weights and Measures Act, 1878, shall be payable to such inspector by the person acting as treasurer of the local authority of the district on presentation of accounts of such expenses, to be furnished quarterly, certified to be correct by the county inspector of the county. *Expenses of ex-officio inspectors. 41 & 42 Vict. c. 49, s. 80.*

The secretary of every grand jury being a local authority under the said Act, shall, at each assizes or presenting term, and the clerk of every other local authority shall, once in every year lay before each such grand jury or other local authority an estimate of the sum which may appear to be necessary to meet such expenses until the next assizes or presenting term, or for the ensuing year; and every such grand jury or other local authority shall, without previous application to the presentment sessions, or other preliminary proceedings, present in advance to the person acting as treasurer the sum specified in such estimate, to be raised and paid out of the local rate; and if the sum so raised proves more than sufficient for the purpose, the balance shall be carried to the credit of the local rate by the person acting as treasurer, and if the sum so raised proves insufficient, the person acting as treasurer shall apply for payment of such expenses any other available funds in his hands.

This provision is new, and appears for the first time in the Act of 1878. It corresponds, however, to the present practice.

Nothing in the Weights and Measures Act, 1878, shall authorise the local authority in Ireland, except the local authority of the borough of Dublin, to appoint inspectors of weights and measures, but such head or other constables in each petty sessions *Ex-officio inspectors of weights and measures. 41 & 42 Vict. c. 49, s. 81.*

district as may be from time to time selected by the inspector general of constabulary, with the approval of the Lord Lieutenant, shall be ex-officio inspectors of weights and measures under that Act within that district, and shall perform their duties under that Act, under the direction of the justices of petty sessions, without fee or reward, and notwithstanding any manorial jurisdiction or claim of jurisdiction within such district.

Provided, that if within one month from the date of such selection the justices signify their disapproval of the selection of any head or other constable, another selection shall be made by the same authority, subject to the same conditions, and the inspector general of constabulary shall, within three days after any selection has been made in a petty sessions district, give or cause to be given to the clerk of that district notice of such selection, and the clerk shall immediately make known the said selection to the justices of the district.

An ex-officio inspector of weights and measures may exercise, without any authority from a justice of the peace, the powers given by the said Act to an inspector of weights and measures having such authority.

In the district in which the commissioners of the Dublin metropolitan police are the local authority under the said Act, such of the superintendents, inspectors, or acting inspectors of the said police as may be selected by the local authority, with the approval of the Lord Lieutenant, shall be ex-officio inspectors of weights and measures within the said district.

This section is made up of parts of several sections in the Acts of 1860-62. The Town Council of Dublin are authorised to appoint inspectors, a power which they previously possessed under the 30 & 31 Vict. c. 94, which has now been repealed.

The local standards of every county or borough in Ireland shall be in the custody of such sub-inspector of constabulary as may be from time to time appointed for that county or borough by the inspector general of constabulary, with the approval of the Lord Lieutenant.

Custody and use of local standards. 41 & 42 Vict. c. 49, s. 82.

Such sub-inspector shall, subject to such regulations as the inspector general of constabulary, with the approval of the Lord Lieutenant, from time to time makes, compare with the local standards in his custody, and adjust and verify the local sub-standards sent to him for the purpose, and when the same are correct shall stamp the same with a stamp of verification, and for the purpose of such verification and stamping, and of the verification of local standards, such sub-inspector of constabulary shall be deemed to be an inspector of weights and measures appointed under the Weights and Measures Act, 1878.

There were similar provisions in the Acts of 1860-62. See 23 & 24 Vict. c. 119, s. 8, and 25 & 26 Vict. c. 76, s. 6.

The local sub-standards shall be deposited in the custody of the ex-officio inspector of weights and measures, and shall at least once in every year, and also at other times when required by the county inspector of constabulary of the county, or by the justices in petty sessions of the county, be compared with the local standards of the county and verified, and when so verified shall, until the expiration of one year or any shorter period at which the next comparison of the same under this section is made, be deemed to be local sub-standards, and be valid local standards for the purpose of the comparison by way of verification or inspection of weights and measures under the Weights and Measures Act, 1878.

Custody and periodical verification of local sub-standards. 41 & 42 Vict. c. 49, s. 83.

The sub-standards provided by the commissioners

of the Dublin metropolitan police shall be verified by comparison with the local standards of the city of Dublin as directed by this section, with this qualification, that the said commissioners, and not the county inspector or the justices, shall have authority to require the same to be verified oftener than once a year.

Any person who uses any sub-standard for any purpose other than that authorised by the said Act shall be liable to a fine not exceeding five pounds.

<small>This section reproduces, in a form in accordance with the rest of the statute, the former provisions in the Acts of 1860–62. See 23 & 24 Vict. c. 119, ss. 8, 18.</small>

<small>Recovery of fines, &c. 41 & 42 Vict. c. 49, s. 84.</small>

For the purpose of the prosecution of offences and the recovery of fines under the Weights and Measures Act, 1878, in Ireland,—

1. The expression "Summary Jurisdiction Acts" in that Act means, within the police district of Dublin metropolis, the Acts regulating the powers and duties of justices of the peace for such district, or of the police of such district, and elsewhere in Ireland the Petty Sessions (Ireland) Act, 1851, and any Act amending or affecting the same; and
2. A court of summary jurisdiction, when hearing and determining an information or complaint in any matter arising under the said Act, shall be constituted within the police district of Dublin metropolis of one of the divisional justices of that district sitting at a police court within the district, and elsewhere of a stipendiary magistrate sitting alone, or with others, or of two or more justices of the peace sitting in petty sessions at a place appointed for holding petty sessions; and

3. Appeals from a court of summary jurisdiction shall lie in the manner and subject to the conditions and regulations prescribed in the twenty-fourth section of the Petty Sessions (Ireland) Act, 1851, and any Acts amending the same.

In the Weights and Measures Act, 1878, unless the context otherwise requires, Definitions. 41 & 42 Vict. c. 49, s. 85.

The expression "Lord Lieutenant" means the lieutenant or other chief governor or governors of Ireland for the time being:

The expression "Treasurer" includes the finance committee, and the secretary of the grand jury for the county of Dublin.

All offences under the Weights and Measures Act, 1878 may be prosecuted, and all fines and forfeitures under that Act may be recovered on summary conviction before a court of summary jurisdiction, in manner prescribed by the Summary Jurisdiction Act (41 & 42 Vict. c. 49, s. 56). By the above section the "Summary Jurisdiction Acts" mean in Ireland the Petty Sessions (Ireland) Act, 1851, for the country generally, and for the police district of Dublin, the local Acts. The Petty Sessions (Ireland) Act, 1851 (14 & 15 Vict. c. 93), consolidated and amended the Acts regulating the proceedings at petty sessions, and the duties of justices of the peace at the Quarter Sessions in Ireland. The provisions of this Act will be found duly described and commented upon in "Humphrey's Justice of the Peace for Ireland." The following short abstract of those provisions of the Act which apply to summonses under the Weights and Measures Act may, however, be useful.

Whenever it is intended that a summons only shall issue to require the attendance of any person, the information or complaint may be made either with or without oath, and either in writing or not, according as the justice shall see fit. But whenever it is intended that a warrant shall issue for the arrest of any person, the information shall be in writing, and on the oath of the complainant, or some person on his behalf. When the information has been thus taken upon oath and in writing, the justice may bind the informant by recognizance to appear and prosecute.

The complaint must be made within six months from the time when the offence shall have been committed, and the defendant is entitled to have a copy of the information when it is in writing. The justice receiving the information may issue his summons requiring the defendant to appear and answer the complaint, and this justice need not be one of those justices by whom the case will be afterwards heard and determined.

If the defendant does not appear at the required time and place, and it is proved, on oath, either that he was *personally* served with the summons, or is keeping out of the way of such service, the justice may issue a warrant to arrest him, and when so arrested, he may either be committed to gaol until the hearing, or discharged, upon entering into a recognizance either with or without sureties at the discretion of the justice.

A summons must be served upon the defendant by delivering to him a copy of the summons, or if he cannot be conveniently met with, by leaving such copy for him at his last or most usual place of abode, or at his office, warehouse, counting-house, shop, factory, or place of business, with some inmate of the house, not being under sixteen years of age, a reasonable time before the hearing of the complaint; and the person serving the summons must indorse upon the same the time and place where it was served, and must attend at the hearing to prove, if necessary, the service.

A justice may enforce the attendance of witnesses by summons, and, if the summons is disobeyed, by warrant. The informer or inspector may give evidence in all cases, and all the witnesses must be examined upon oath.

When the defendant appears in answer to the summons the complaint is stated to him, and if he admits the truth of the same, the justices, if they see no sufficient reason to the contrary, convict him. If he does not admit the truth, then evidence is adduced in support of the complaint. This is followed by evidence on behalf of the defence, which the complainant may meet with evidence in reply, if the defendant's evidence is other than as to his general character. The complainant or his agent cannot make a speech in reply to the evidence for the defence, nor can the defendant make a second speech after the evidence in reply.

If the defendant does not appear at the hearing, and the service of the summons is duly proved, the justices may either hear and determine the case in the defendant's absence, or may adjourn the hearing to a future day.

If the defendant appears, and the complainant does not, the justices may either dismiss the case or adjourn the hearing.

When required to do so by either party or his agent, the justices shall take or cause to be taken a note in writing of the evidence, or of so much as is material, in a book to be kept for that purpose by their clerk, which book shall be signed by one of the justices by whom the complaint has been heard upon the day on which the same has been determined. The justices may, in their discretion, adjourn the hearing of a case to a certain time or place to be appointed and stated in the presence and hearing of the parties or their agents, and may either allow the defendant to go at large, or (where the information has been on oath and in writing) may commit him to gaol by warrant, or may discharge him upon his entering into a recognizance, either with or without sureties, to appear at the adjournment.

The justices having heard what each party has to say, and the evidence on both sides, will either convict or dismiss the complaint on the merits, or without prejudice to its being again made, and the result is duly entered in the "Order Book," and signed by one of the justices present.

The justices have several general powers in adjudicating upon cases under their summary jurisdiction. They may order the penalties to be paid forthwith, or may give time for payment, and may order distress upon non-payment of the penalty. In default of distress they may adjudge imprisonment upon the following scale :—For any sum exceeding five pounds but not exceeding ten pounds, the imprisonment not to exceed three months; exceeding ten pounds but not exceeding thirty pounds, four months; exceeding thirty pounds but not exceeding fifty pounds, six months; exceeding fifty pounds, one year. Such imprisonment is determinable upon payment of the fine and costs, and costs of the distress where the same has been made. Where a person has no goods, or a distress would be ruinous to him, the justices may order imprisonment in the first instance; and where the penalty is less than five pounds the following terms of imprisonment, by the "Small Penalties (Ireland) Act, 1873," may be adjudged without any warrant of distress. For any penalty not exceeding ten shillings, the imprisonment not to exceed seven days; exceeding ten shillings but not exceeding one pound, fourteen days; exceeding one pound but not exceeding two pounds, one month; exceeding two pounds but not exceeding five pounds, two months. See 36 & 37 Vict. c. 82.

The justices have power to substitute distress for committal and *vice versâ* on the failure of the first warrant; but in cases under the Weights and Measures Act there is no power to award imprisonment with hard labour. They have also power to award costs to the defendant, if the case is dismissed, and to the complainant if there is a conviction, but the sum awarded is not to exceed twenty shillings. Aiders and abettors in the commission of any offence can be punished upon summary conviction in the same way as their principals.

In all cases of summary jurisdiction, whenever an order is made upon the conviction of any person for an offence, the justice issues the proper warrant for its execution; but when the party, being entitled to appeal against any such order, gives notice thereof and enters into a recognizance to prosecute the same, the justice is not to issue any warrant to execute the order until the appeal is decided, or the appellant has failed to perform the condition of his recognizance as the case may be. If the warrant has been executed and the appellant is in custody or has been committed to gaol, or if a warrant of distress has been issued or executed, under any such order, the justice by whom the warrant has been issued, or any other justice of the same county, upon application being made to him on that behalf, shall forthwith order the discharge of such person from custody or from gaol, or that the warrant of distress is not to be executed, or, if it has been executed, that the distress shall be returned to the owner, as the case may be.

Proceedings upon appeal.—"Appeals from a court of summary

jurisdiction shall lie in the manner and subject to the conditions and regulations prescribed in the twenty-fourth section of the Petty Sessions (Ireland) Act, 1851, and any Acts amending the same." This 24th section provides that where an order is made by parties for payment of any penal or other sum exceeding twenty shillings, or for the estreating of any recognizance to a greater amount than twenty shillings, the party against whom the order is made is entitled to appeal to the next Quarter Sessions to be held in the same division of the county, where the order has been made by the justices of any petty sessions district, or to the recorder of any corporate or borough town at his next sessions, where the order has been made by the justices of such corporate or borough town, unless such sessions commence within seven days from the date of such order, in which case the appeal may be made to the next succeeding sessions of such division or town.

The appeal is subject to the following provisions:—

The appellant must serve notice in writing of his intention to appeal upon the clerk of petty sessions, within three days from the date of the order against which the appeal is made. Within three days after such notice he must enter into a recognizance with two solvent sureties, conditioned to prosecute the appeal, and the amount of the recognizance is to be double the amount of the sum and costs ordered to be paid. When these have been done, the appellant is to receive a certificate of the order against which he appeals, and a certificate that he has duly given notice and entered into a recognizance, if these have been done. The clerk of petty sessions has then to transmit the recognizance and all other proceedings in the case to the clerk of the peace of the county, or to the proper officer of the Recorder's court, at least seven days before the commencement of the sessions to which the appeal is made, or as soon after as may be practicable. The appellant must also give notice in writing to the opposite party of his intention to prosecute his appeal at least seven clear days before the commencement of the sessions to which the appeal is made.

Whenever the appeal has been made, and the notice to the opposite party been duly given, the Court of Quarter Sessions or the Recorder, as the case may be, may entertain the same, and may confirm, vary, or reverse the order made by the justices, and may award to either party any sum not exceeding forty shillings for the costs of such appeal. When the appeal has been decided, the officer of the court certifies such decision as part of the form of appeal, and returns the same and the said proceedings to the justices of the petty sessions within seven days after the appeal has been decided; and if the appeal has not been duly prosecuted, the said officer of the court certifies the same upon the recognizance which, within seven days after the termination of the sessions at which the appeal ought to have been prosecuted, he transmits to the justices of the petty sessions. When it appears that the appeal has not been duly prosecuted, or that the order has been confirmed upon appeal, a proper warrant for the execution of the original order may be issued as if no appeal had been brought. When the order has been varied, the

Scotland and Ireland. 113

warrant must be issued for the execution of the order as varied by the court of appeal; and the costs of the appeal may be enforced in the same manner as the costs awarded by the original order.

Whenever the party bound to prosecute the appeal has no goods whereon to levy by distress, the justices at the petty sessions where the original order was made, and after due proof of notice to the parties, may estreat the recognizance to such amount as they may see fit, and for paying out of such amount the sum directed to be paid to any party by the original order, and may issue a warrant for the levy of the same upon the goods of the several persons bound thereby.

There are also provisions in the Petty Sessions (Ireland) Act, 1851, for addressing and executing warrants, and in a schedule will be found forms applicable to all cases, which are to be deemed valid and proper forms in all proceedings.

For further details see the Act itself (14 & 15 Vict. c. 93).

In Ireland, in every city or town, not being a county of itself, every person, persons, or body corporate exercising the privilege of appointing a weigh-master, shall supply him with accurate scales, and with an accurate set of copies of the local standards, and in default shall be liable on summary conviction to a fine of twenty pounds, and the accuracy of such set of copies shall be certified under the hand of some inspector of weights and measures. They shall also, once at least in every five years, cause such copies to be readjusted by comparison with some local standards which have been verified by the Board of Trade, and in default, shall be liable on summary conviction to a fine of five pounds. Supply of weigh-masters in Ireland with scales and copies of local standards. 41 & 42 Vict. c. 49, schedule.

Such set of copies shall, for the purpose of comparison and verification, be considered local standards, and shall be used for no other purpose whatever, and if they are so used, the person using the same shall be liable on summary conviction to a fine of five pounds.

This section, which is in the second part of the sixth schedule of the Act of 1878, is, with some slight necessary changes, a re-enactment of section 26 of the Weights and Measures Act, 1835. It has been inserted in a schedule of the Act of 1878, so that when it is hereafter consolidated with the Acts to which it will more properly belong it may be struck out, without altering a material part of the Act in which it has now, for the present, been placed.

I

The following unrepealed sections of the "Weights and Measures (Ireland) Amendment Act, 1862," are still in force.

Penalty on counterfeiting of brand.
25 & 26 Vict. c. 76, s. 14.

If any person commit any of the following offences, he shall for each offence be liable to a penalty not exceeding five pounds:—

1. If he, with intent to defraud, counterfeit or procure to be counterfeited any brand or stamp used by or under the authority of the owner or lessee of a market or fair, or of any person having by law the control of a market or fair, to denote the weight, measure, or quality of any article sold in the market or fair, or within the prescribed limits, during the holding of the market or fair, or of any cask, firkin, or other vessel, covering, or thing in which such article is sold, or the impression of any such brand or stamp;
2. Or, with the like intent, use or procure to be used any such counterfeit brand, stamp, or impression;
3. Or, with the like intent, alter an impression of any such genuine brand or stamp;
4. Or, with the like intent, have in his possession anything having thereon an impression of any such counterfeit brand or stamp, or a fraudulently altered impression of any such genuine brand or stamp;
5. Or, with the like intent, transfer or apply any cask, firkin, or other vessel, covering, or thing, having thereon an impression of any such genuine brand or stamp, to any article other than that for denoting the weight, measure, or quality whereof such impression was made on such cask, firkin, or other vessel, covering, or thing; or in any other manner alter the bonâ fide application of an impression of any such genuine brand or stamp;

6. Or, knowingly weigh or cause to be weighed, contrary to the provisions of the Weights and Measures Act, 1878, or act or assist in committing, or connive at any fraud respecting the weighing, or the weight or measure of any such article as in section 76 of the Weights and Measures Act, 1878, is mentioned;
7. Or, with intent to defraud or alter any ticket specifying the weight of any such article;
8. Or, with intent to defraud, make, or use, or be privy to the making or using of any such ticket, falsely stating the weight of any such article, or of any covering, cart, or load;
9. Or, shall dispose of, sell, or cause to be sold, any weight or measure having a false or counterfeit stamp, or a stamp purporting to resemble a genuine stamp.

If any person shall wilfully pack up or mix, or cause to be packed up or mixed, with or in any butter contained in any firkin or cask, any salt, pickle, or other substance, with intent to increase the weight of such butter, and shall bring or send any butter so packed or mixed to any market for sale, he shall be liable to pay a fine not exceeding forty shillings, or be imprisoned for any period not exceeding one month, as the justice or justices shall determine. *Penalty for fraudulently increasing weight of butter in casks.* 25 & 26 Vict. c. 76, s. 15.

If any person shall wind or cause to be wound in any fleece any wool not being sufficiently rivered or washed, or wind or cause to be wound within any fleeces any deceitful locks, cots, skin, or lamb's wool, or any substance, matter, or thing whereby the fleece may be rendered more weighty, to the deceit and loss of the buyer, such person shall be liable to *Penalty for fraudulently increasing weight of fleeces.* 25 & 26 Vict. c. 76, s. 16.

a penalty of two shillings for every fleece so fraudulently made up.

Penalties, how recoverable. 25 & 26 Vict. c. 76, s. 17.
Any penalty recoverable under the provisions of this Act, shall be recoverable in a summary way, with respect to the police district of Dublin metropolis, subject and according to the provisions of any Act regulating the powers and duties of justices of the peace for such district, or of the police of such district, and with respect to other parts of Ireland, before a justice or justices of the peace sitting in petty sessions, subject and according to the provisions of "The Petty Sessions (Ireland) Act, 1851," and any Act amending the same.

Limitation of proceedings for penalties. 25 & 26 Vict. c. 76, s. 18.
No penalty imposed by this Act shall be recovered unless proceedings for recovery thereof are commenced within three months next after the commission of the alleged offence, or in case of a continuing offence within three months after the alleged offence has ceased to be committed.

Nothing to prevent persons being indicted for offences. 25 & 26 Vict. c. 76, s. 19.
Nothing in this Act shall be taken to prevent any person from being indicted for any indictable offence made punishable on summary conviction by this Act, or to prevent any person from being liable under any other Act or otherwise to any other or higher penalty or punishment than is provided for any offence by this Act, but so that no person be punished twice for the same offence.

CHAPTER XII.

THE SALE OF COALS AND BREAD.

(1.) *The Sale of Coals.*

ALL coals, slack, culm, and cannel of every description shall be sold by weight, and not by measure. Every person who sells any coals, slack, culm, or cannel of any description by measure, and not by weight, shall be liable on summary conviction to a fine not exceeding forty shillings for every such sale. <small>Sale of coals by weight and not by measure. 41 & 42 Vict. c. 49, sched.</small>

The sale of coals by weight and not by measure is compulsory throughout the United Kingdom. The above section is a re-enactment of a similar provision in the Weights and Measures Act, 1835 (5 & 6 Will. IV. c. 63, s. 9).

The two following cases have been decided with reference to this section.

Where, by a Navigation Act, certain rates and duties were imposed on coals, &c., landed within a certain district, to be paid to commissioners therein named; after the passing of the Act, 5 & 6 Will. IV. c. 63, the commissioners had power to levy the rates by the *ton*, they having been previously levied by the *chaldron*, without first applying to the sessions for an inquisition under section 14 of the same Act (*Goody* v. *Penny*, 9 M. & W. 687).

The right of a Corporation by custom by means of their deputies to *measure* all coals imported into the port was not converted into a right to *weigh* them by the Weights and Measures Act, 1835 (*Smith* v. *Cartwright*, 20 L. J., Exch. 401; 6 Exch. R. 927; 17 L.T. 258; 15 J. P. 564).

In 1831 was passed an "Act for regulating the vend and delivery of coals in the Cities of London and Westminster, and in certain parts of the counties of Middlesex, Surrey, Kent, Essex, Hertfordshire, Buckinghamshire, and Berkshire" (1 & 2 Will. IV. c. 76, loc.). This Act was amended in several important particulars in 1838 by the 1 & 2 Vict. c. 101 (loc.); and the regulations for the vend and delivery of coals, and the collection of the coal duties in London, have been the subject of much legislation since that time. These Acts are only in force within the Cities of London and Westminster, and within the distance of twenty-five miles from the

General Post Office in the City of London; but the district is such an important one, that it may be useful to insert some of the more important provisions of the Acts in question.

<small>For preventing the sale of one sort of coals for another. 1 & 2 Will. IV. c. 76, s. 45 (*loc.*).</small> If any seller or sellers of or dealer or dealers in coals shall knowingly sell one sort of coals for and as a sort which they really are not, every such seller or sellers of or dealer or dealers in coals, shall forfeit and pay for every such offence the sum of ten pounds per ton for every ton of coals so sold, and so in proportion for any smaller quantity. Provided always, that no seller or sellers of or dealer or dealers in coals shall be subject to such penalty for or in respect of any number of tons exceeding twenty-five tons for the same offence.

A corporate body such as Guardians of the Poor cannot sue as common informers for penalties unless expressly empowered to do so by the statutes imposing the penalties. As there is no such power given in this Coal Act, it was held that a Board of Guardians could not sue for penalties under the above section (*Shoreditch Guardians* v. *Franklin*, 42 J. P. 727).

<small>Seller's ticket to be sent with coals. 1 & 2 Vict. c. 101, s. 3 (*loc.*).</small> With any quantity of coals exceeding five hundred and sixty pounds, delivered by any cart, waggon, or other carriage, the seller or sellers thereof shall deliver or cause to be delivered to the purchaser or purchasers thereof, or to his, her, or their agent or agents, or servant or servants, immediately on the arrival of the cart, waggon, or other carriage in which such coals shall be sent, and before any of such coals shall be unloaded, a paper or ticket, according to the following form:—

> Mr. A. B. (*here insert the name of the buyer*), take notice, that you are to receive herewith (*here insert the number*) tons (*here insert the name of the coal, if any particular sort is ordered or contracted for as Wall's End, specify the name of the colliery*) coals in (*here insert the number*) sacks, containing (*here insert the weight*) pounds of coal in each sack.
>
> Signed C. D. (*here insert the name or names of the seller or sellers in words at full length*).
>
> E. F. (*here insert the name of the carman in words at full length*).

The Sale of Coals and Bread. 119

It is directed that with any quantity of coals exceeding five hundred and sixty pounds, a paper or ticket describing the quantity, and if any particular sort is ordered or contracted for the sort of the coals sent by the seller, shall be delivered to the purchaser, or his agent or servant, before any part of such coals shall be unloaded; that a weighing machine or proper scales and weights shall be carried with every waggon, cart, or other carriage, and the carman is required to weigh gratuitously any sack or sacks of coal which shall be chosen by the purchaser or his agent or servant; and if any carman refuses to weigh such sack or sacks of coals as aforesaid, or drives away the waggon, cart, or other carriage, before the coals are weighed, or otherwise obstructs the weighing thereof, he is liable to a penalty not exceeding twenty pounds; also, that a proper machine or proper scales and weights for weighing coals shall be kept at every watch-house or police station, and at any other place appointed for that purpose by two or more of Her Majesty's justices of the peace.

And in case any such seller or sellers do not deliver or cause to be delivered such paper or ticket as aforesaid to the purchaser or purchasers of such coals, or to his, her, or their agent or agents, or servant or servants, before any part of such coals are unloaded, every such seller shall for every such offence forfeit and pay any sum not exceeding twenty pounds; and in case the carman, driver of, or other person attending any such cart, waggon, or other carriage laden with any such coals, to whom any such paper or ticket shall have been given by, or by the orders of the seller in order to be delivered to the purchaser, shall (having so first received the same from the seller or any person by the direction of the seller) refuse or neglect to deliver such paper or ticket to the purchaser or purchasers of such coals, or to his, her, or their agent or agents, or servant or servants, before any part of such coals shall be unloaded, such carman, driver, or other person so offending shall for every such offence forfeit and pay any sum not exceeding twenty pounds: Provided always, that coals delivered to any seller or dealer in coals, or to any person or persons purchasing the same at the Coal Market, may be delivered without any such paper or ticket.

The omission to deliver a ticket in accordance with the provisions of this section is a bar to an action for the sale and delivery of coals exceeding in quantity five hundred and sixty pounds (*Cundell* v. *Dawson*, 4 C. B. 376; 17 L. J. C. P. 311; *Little* v. *Poole*, 9 B. & C. 192).

This section, which follows one repealing so much of the Act of 1831 as related to the delivery of a ticket, does not provide for recovering the penalty by any individual, so that no action can be maintained for this penalty by the buyer of coals, where no ticket is delivered (*Meredith* v. *Holman*, 16 M. & W. 798; 16 L. J. Ex. 126).

<small>A paper or ticket to be sent with coals delivered from lighters.
1 & 2 Vict. c. 101, s. 4 (*loc.*).</small>

With any quantity of coals exceeding five hundred and sixty pounds, delivered, by any lighter, vessel, barge, or other craft, the seller or sellers thereof shall deliver or cause to be delivered to the purchaser or purchasers thereof, or to his, her, or their agent or agents, or servant or servants, immediately on the arrival of the lighter, vessel, barge, or other craft in which such coals shall be sent, and before any of such coals shall be unloaded, a paper or ticket setting forth in words at length the number of tons to be delivered, and the name of the coals, together with the name and number of the lighter, vessel, barge, or other craft, and the name of the seller or sellers, and also the name of the lighterman; and in case any such seller or sellers do not deliver or cause to be delivered such paper or ticket as aforesaid to the purchaser or purchasers of such coals, or to his, her, or their agent or agents, or servant or servants, before any part of such coals is unloaded, every such seller shall for every such offence forfeit and pay any sum not exceeding twenty pounds; and in case the person having the charge of the lighter, vessel, barge, or craft laden with any such coals, to whom any such paper or ticket should have been given by or by the order of the seller, in order to be delivered to the purchaser, shall (having so first received the same from the seller, or any person by the direction of the seller) refuse or neglect to deliver such paper or ticket to the purchaser or purchasers of such coals, or to his, her, or their agent or agents, or servant or servants, before any part of such coals shall be unloaded, the person so offending shall for every such offence forfeit and pay any sum not exceeding twenty pounds. Provided always, that coals delivered to any seller or dealer in coals, or any person or persons purchasing the same at the Coal Market, may be delivered without any such paper or ticket.

<small>The delivery of coals to a purchaser direct out of the vendor's coal brig on to the purchaser's wharf, within the district, without the intervention of any lighter, barge, or other craft, is not a delivery from a "vessel" within the limited meaning of the word in the above section, and therefore does not require to be accompanied by a ticket (*Blandford* v. *Morrison*, 19 L. J. Q. B. 533).</small>

All coals sold from any lighter, barge, or other craft, or from any wharf, warehouse, or other place, in any quantity exceeding five hundred and sixty pounds, except coals carried and delivered in bulk as hereinafter mentioned, shall be carried and delivered to the respective purchasers thereof in sacks, each sack containing either one hundred and twelve pounds or two hundred and twenty-four pounds net: Provided always, that coals, delivered by gang labour may be conveyed in sacks containing any weight, anything herein contained to the contrary thereof notwithstanding. *Coals to be delivered in sacks containing a certain quantity. 1 & 2 Will. IV. c. 76, s. 48 (loc.).*

Any coals sold from any ship, lighter, barge, or other craft, or from any wharf or place, in any quantity exceeding five hundred and sixty pounds, may be carried and delivered to the respective purchasers thereof, if they think fit, in bulk, in carts or other carriages, or in any lighter, barge, or other craft: Provided also, that in every case where any such coals shall be carried and delivered in any cart or other carriage, in bulk as aforesaid, the weight of such cart or other carriage, as well as of the coals contained therein, shall be previously ascertained by a weighing machine fixed for that purpose on the wharf or place from which the coals *Coals may be delivered in bulk. 1 & 2 Will. IV. c. 76, s. 49 (loc.).*

The Sale of Coals and Bread.

shall be brought ; and the seller's ticket shall in such cases state the weight of the cart or other carriage, as well as the weight of the coals contained therein ; and if any sellers or dealers in coals shall carry or deliver to the purchaser or purchasers by any cart or other carriage, any quantity of coals exceeding five hundred and sixty pounds in bulk, without having a weighing machine fixed up on his wharf or place, or without having previously ascertained by such weighing machine the weight of the cart or other carriage, and the weight of the coals contained therein, then and in every such case such seller or dealer shall for every such offence forfeit and pay any sum not exceeding fifty pounds.

Carman to weigh the carriage and the coals, if required
1 & 2 Will. IV. c. 76, s. 50 (loc.).

The carman or driver of any cart, waggon, or other carriage in which any coals exceeding in quantity five hundred and sixty pounds shall be carried in bulk, for delivery to the purchaser or purchasers thereof, from any ship, lighter, barge, or other craft, or from any wharf, warehouse, or other place shall (in case he shall be required so to do by the purchaser or purchasers of such coals, or his, her, or their servant or servants, or other person or persons acting on the behalf of such purchaser or purchasers) weigh the waggon or other carriage, with the coals therein, at any public weighing machine for carts or carriages, which may be situate on the road between the place from which the coals shall be brought and the place of delivery, or at any point within the distance of one hundred yards from any part of such road ; and such carman or driver is also hereby directed (in case he shall be required so to do by the purchaser or purchasers, or any such other person or persons as aforesaid) to weigh in like manner the cart, waggon, or other carriage, without the coals, at any public weighing machine for carts or carriages which may be situate as aforesaid : and if any such carman or driver shall neglect or refuse, when so required as aforesaid, to weigh the cart, waggon, or other carriage, either with or without the coals, at any public weighing machine for carts and carriages which may be situate as aforesaid, such carman or driver shall for every such offence forfeit any sum not exceeding ten pounds: Provided always, that no carman or driver shall be compelled or obliged to weigh the cart, waggon, or other carriage without the coals, until after the same shall have been delivered, and that no such carman or driver shall be obliged to go back or return to any such public weighing machine as aforesaid, for the purpose of weighing the cart, waggon, or other carriage, either with or without the coals, after he shall have passed the same.

If in any case where any coals shall be delivered in bulk to the purchaser or purchasers thereof, from any ship, lighter, barge, or other craft, or from any wharf, warehouse, or other place a less quantity shall be delivered than shall be expressed in the ticket to be delivered therewith as aforesaid, the seller or sellers shall for every such offence forfeit any sum not exceeding ten pounds; and if the deficiency shall exceed two hundred and twenty-four pounds, the seller or sellers shall forfeit any sum not exceeding fifty pounds. Penalty on deficiency in weight of coals. 1 & 2 Will. IV. c. 76, s. 51 (*loc.*).

If any carman or driver of any cart, waggon, or other carriage laden with coals for sale, or to be delivered to the purchaser or purchasers thereof, by any seller or sellers of, or dealer or dealers in, or carrier or carriers of coals, from any ship, lighter, barge, or other craft, or from any wharf, warehouse, or other place shall not have placed in, on, or under his cart, waggon, or carriage a perfect weighing machine marked at *Guildhall, London*, by the proper officer there, for which the sum of two shillings and sixpence shall be paid, and no more (which machine shall be of the form, size, and dimensions of the Carman to carry a weighing-machine in his cart. 1 & 2 Will. IV. c. 76, s. 52 (*loc.*).

machine approved by the Lord High Treasurer, or any three or more of the Lord Commissioners of Her Majesty's Treasury, and deposited at the office of the Hall Keeper of the City of London, to which any person shall have access between the hours of ten in the morning and two in the afternoon, and shall be provided by the seller or sellers, dealer or dealers in, or carrier or carriers of such coals) then and in every such case every such carman or driver of such cart, waggon, or other carriage not having such machine so placed therein, thereon, or thereunder, shall for every such offence forfeit and pay any sum not exceeding ten pounds, and the seller or sellers of, or dealer or dealers in, or carrier or carriers of, such coals shall forfeit and pay any sum not exceeding twenty pounds : Provided always, that coals which shall be carried or conveyed in bulk, or in any cart, waggon, or any other carriage belonging to the purchaser or purchasers of such coals, may be so carried or conveyed without the carman being obliged to carry a weighing machine therewith or any person or persons being subject or liable to any penalty or penalties in respect thereof.

As to weighing machines required to be sent with each cart. 1 & 2 Vict. c. 101, s. 5 (loc.).

No weighing machine shall be deemed a perfect weighing machine within the meaning of the foregoing section unless proper weights shall be carried therewith, and any other just balance, with an even beam and legal weights, shall be deemed a perfect weighing machine within the meaning of the said section, without having been marked at *Guildhall;* and if any carman or driver required to carry a weighing machine shall have placed in, on, or under his cart, waggon, or other carriage, any beam, or scales, or other weighing machine, or any weights which shall be imperfect or improper for the purpose of denoting the weight of coals, then, and in every such case every such carman or driver, or person delivering such coals, shall, for every such offence, forfeit and pay any sum not exceeding five pounds, and the seller or sellers of, or dealer or dealers in, or carrier or carriers of, such coals shall forfeit and pay any sum not exceeding ten pounds.

The carman or driver of any cart, waggon, or other carriage in which coals shall be carried in sacks for delivery to the purchaser or purchasers thereof, from any ship, lighter, barge, or other craft, or from any wharf, warehouse, or other place, shall, and he is hereby directed to weigh, if he shall be required so to do, any one or more of the sacks contained in any such cart, waggon, or other carriage which may be chosen by the purchaser or purchasers of the said coals, or his, her, or their servant or servants, or other person or persons acting on the behalf of such purchasers, with the coals therein, and also afterwards to weigh in like manner such sack without any coals therein. *Carman required to weigh any of the sacks in the cart.* 1 & 2 Will. IV. c. 76, s. 54 (*loc.*).

From any ship, lighter, barge, &c.—In an action for penalties under the Coal Act, the coals must have been delivered *from* a ship, wharf, &c., within the prescribed limits, and not merely delivered *at* a place within the district to which the Act applies (*Frend* v. *Butterfield*, 11 A. & E. 838).

With the coals therein, and also afterwards to weigh in like manner such sack without any coals therein.—To weigh each sack of coals in one scale against weights in the other scale equal to the proper weight of a sack of coals, together with an empty sack, is not a legal weighing within this section (*Meredith* v. *Holman*, 16 M. & W. 798; 16 L. J. Ex. 126).

If any carman or driver of any cart, or waggon, or other carriage in which coals shall be carried in sacks for delivery to the purchaser or purchasers thereof, from any ship, vessel, lighter, barge, or other craft, or from any wharf, warehouse, or other place, shall neglect or refuse to weigh by the said machine any such sack or sacks of coal in manner hereinbefore directed, when thereunto required by the purchaser or purchasers of such coals, or by his, her, or their servant or servants, or other person or persons acting by, for, or under the authority of such purchaser or pur- *Penalty on carman for driving coals away without weighing if required.* 1 & 2 Will. IV. c. 76, s. 55 (*loc.*); and 1 & 2 Vict. c. 101, s. 6 (*loc.*).

chasers, or if any such carman or driver shall drive away, or permit, or suffer the said cart, waggon, or other carriage to be driven away, without weighing in manner herein directed the said sack or sacks of coals, or shall hinder, obstruct, or otherwise prevent the purchaser or purchasers of such coals, or his, her, or their servant, from examining the said machine, or weighing all or any of the sack or sacks of coals in such his cart, waggon, or other carriage, then and in every such case every such carman or driver so offending shall for every such offence forfeit and pay any sum not exceeding twenty pounds nor less than five pounds.

All the coals sent to be weighed if desired by the purchaser.
1 & 2 Will. IV. c. 76, s. 56 (loc.).

If any purchaser or purchasers, or his, her, or their servant or servants, or any other person or persons acting by, for, or under the authority of such purchaser or purchasers who shall require any sack or sacks of coals to be weighed as aforesaid, shall find the coals therein to be deficient in weight, and shall signify to the carman or other person attending such cart, waggon, or other carriage, his, her, or their desire to have all the coals contained in such cart, waggon, or other carriage, or any part of such coals, weighed or reweighed in the presence of some constable, police officer, or other indifferent and credible person, then, and in every such case the carman or driver of such cart, waggon, or other carriage in which such coals shall be brought shall, and he is hereby required to continue and remain at or before the house, lodging, or other premises of the purchaser or purchasers of such coals, with such coals, and the cart, waggon, or other carriage, until such coals are weighed; and if any such carman or driver shall drive away, or permit or suffer to be driven away, such cart, waggon, or other carriage, before the coals contained therein shall be weighed, without the consent of the purchaser or purchasers thereof, or his, her, or their servant or servants, or such other person or

The Sale of Coals and Bread.

persons as aforesaid, then, and in every such case such carman or driver shall, for every such offence, forfeit and pay any sum not exceeding twenty pounds.

Such purchaser or purchasers, or his, her, or their servant or servants, or other person or persons as aforesaid, so desiring such coals contained in such cart, waggon, or other carriage to be weighed, shall, and he, she, or they is and are required to procure the attendance of some constable, police officer, or other indifferent and credible person, to be present at the weighing of such coals; and all the said sacks, both with and without the coals therein, shall accordingly be weighed with the said machine by the carman or other person attending such cart, waggon, or other carriage, in the presence of the purchaser or purchasers of the said coals, or of his, her, or their agent or servant, agents or servants, if they or any of them shall attend to see the same weighed, and of such constable, police officer, or other person; and in case such purchaser or purchasers, or his, her, or their agent or servant, agents or servants, shall not attend for the purpose of seeing such coals so weighed, then such carman or other person shall proceed in the weighing of such sacks in his, her, or their absence; and in case such carman or other person shall refuse or neglect to weigh such sacks, or any of them, in manner aforesaid, he shall forfeit and pay for such offence any sum not exceeding ten pounds; and the constable, police officer, or any other person who may be present may weigh the said sacks or any of them as aforesaid; and in case upon the weighing of any such sack or sacks it shall happen that any sack or sacks shall not contain either one hundred and twelve pounds or two hundred and twenty-four pounds net of coals, as the case may be, then, and in every such case the seller or sellers of such coals shall for every such sack of coals that shall be so found deficient forfeit and pay any sum not exceeding five pounds.

Purchaser to procure the attendance of a constable, &c., if desirous of having the coals re-weighed. 1 & 2 Will. IV. c. 76, s. 57 (loc.).

Where several sacks are sent out to a purchaser at the same time under one contract, one penalty only is incurred in respect of a deficiency in weight, though every sack is so deficient. The penalty incurred is five pounds for each sack and if the amount exceeds £25 it must be recovered by action and not before justices of the peace, as provided in other sections of the Act (*Collins* v. *Hopwood*, 15 M. & W. 459; 16 L. J. Ex. 124).

<small>No quantity less than 560 pounds weight of coals to be sold without being weighed. 1 & 2 Will. IV. c. 76, s. 58 (*loc.*).</small>

All coals sold in any quantity less than five hundred and sixty pounds, or in the quantity of five hundred and sixty pounds, from any place, or from any cart or other carriage, shall be weighed previous to being delivered to the purchaser or purchasers of such coals, and also, if required by such purchaser or purchasers, or his, her, or their agent or servant, in the presence of such purchasers, or his, her, or their agent or servant; and if any seller or dealer in coals shall deliver to the purchaser or purchasers thereof any quantity of coals less than five hundred and sixty pounds, or the quantity of five hundred and sixty pounds, without previously weighing the same, and also, if required by such purchaser or purchasers, or his, her, or their agent or servant, in the presence of such purchaser or purchasers, or his, her, or their agent or servant, then, and in every such case such seller or dealer shall for every such offence forfeit and pay any sum not exceeding five pounds.

<small>Weighing machines to be kept at watch-houses and police-stations. 1 & 2 Will. IV. c. 76, s. 59 (*loc.*).</small>

A proper machine or proper scales and weights for weighing coals shall be kept at every watch-house or police station, or at any other place or places which shall from time to time be appointed by any two or more of her Majesty's justices of the peace for the said cities or counties, within the Cities of London and Westminster, or within the distance of twenty-five miles from the General Post-office; and the same shall be provided and kept in repair from time to time by the overseers of the poor of the township, parish, precinct, or place in which such watch-house, station, or such other place or places as aforesaid shall be situate, out of the rate for the relief of the poor of such township, parish, precinct, or place, and shall and may be used at any time or times for weighing, in such township, parish, precinct, or place, any coals respecting which there may be any dispute; and in case the overseers of any such township, parish, precinct, or place shall not provide and send to such watch-house, station, or such other place or places as aforesaid, such a machine, or if such overseers shall not cause such machine to be repaired or a new machine to be provided within seven days after notice of the want thereof in writing shall have been given to them, or left at their usual places of abode, by any police officer, or any inhabitant of such township, parish, precinct, or place, such overseers shall for every such offence forfeit and pay any sum not exceeding ten pounds.

The Acts also provide for the recovery and application of fines and penalties, and the expenses of witnesses, &c.

(2.) *The Sale of Bread.*

In 1822 an Act was passed to provide regulations for the sale of bread. This Act applied only to the "City of London and the liberties thereof, and within the weekly bills of mortality and ten miles of the Royal Exchange" (3 Geo. IV. c. 106, *loc.*).

In 1836 a general Act was passed to provide regulations for the sale of bread without the limits assigned in the Act of 1822; but not to extend to Ireland (6 & 7 Will. IV. c. 37).

As the sections with reference to the sale of bread in the two Acts are almost exactly the same, those from the latter Act are here given as practically applicable to the whole of England and Scotland.

It shall and may be lawful for all bakers or sellers of bread to make and sell, or offer for sale, in his, her, or their shop, or to deliver to his, her, or their customer or customers, bread made of such weight or size as such bakers or sellers of bread shall think fit; any law or usage to the contrary notwithstanding. *Bakers to make bread of any weight or size. 6 & 7 Will. IV. c. 37, s. 3.*

The Sale of Coals and Bread.

Bread to be sold by weight, and in no other manner, under penalty not exceeding forty shillings.
6 & 7 Will. IV. 37, s. 4.

All bread shall be sold by bakers or sellers of bread by weight; and in case any baker or seller of bread shall sell or cause to be sold bread in any other manner than by weight, then and in such case every such baker or seller of bread shall, for every such offence, forfeit and pay any sum not exceeding forty shillings, which the magistrate or magistrates, justice or justices, before whom such offender or offenders shall be convicted, shall order and direct: Provided always, that nothing in this Act contained shall extend or be construed to extend to prevent or hinder any such baker or seller of bread from selling bread usually sold under the denomination of French or fancy bread, or rolls, without previously weighing the same.

The following important cases have been decided with reference to this section.

A baker was in the habit of weighing out dough for two-pound, four-pound, and eight-pound loaves before putting them into the oven, allowing five ounces for shrinkage of a four-pound loaf, which is the understood weight of a "quartern" loaf, but he did not weigh the loaves afterwards unless required by the customer. A customer bought a "quartern" loaf, and the current price of a four-pound loaf was asked and paid. He did not require to have the loaf weighed, and it never was weighed after it was baked; it turned out two ounces nine drachms short of four pounds. Held that the baker was rightly convicted (*Jones* v. *Huxtable*, L. R. 2 Q. B. 460; 8 B. & S. 433; 16 L. T. N.S. 381; 36 L. J. M. C. 122; 31 J. P. 534; and see *Williams* v. *Deggan*, 16 L. T. N.S. 492; 31 J. P. 807; and *Milton* v. *Troke*, 20 L. T. N.S. 563; 33 J. P. 821).

Usually sold under the denomination of French or fancy bread.—These words refer to bread which is usually sold as fancy bread at the time of sale, and not that which was sold as fancy bread at the time of the passing of the Act. Bread which was usually sold as fancy bread in 1836 but which is not now so sold does not come within the proviso (*R.* v. *William Wood*, L. R. 4 Q. B. 599; 10 B. & S. 534; 38 L. J. M. C. 144; 20 L. T. N.S. 654; 33 J. P. 823).

A customer went into a baker's shop and asked for a four-pound loaf, and a loaf was handed to him by the baker, which turned out substantially deficient in weight. The purchaser did not ask to have the loaf weighed, and there was no evidence as to whether the loaf had ever been weighed or not, but the baker contended that the loaf was fancy bread. It was held that the customer, having asked for bread by weight, the baker was bound to sell by weight, whether the bread was ordinary or fancy bread; and though he was not bound to weigh in the presence of the customer, unless requested to do so, he was bound to weigh at some time or other, and that, as the loaf was substantially deficient in weight, it must be taken, as against the baker, that he had never weighed it. The conviction was upheld (*R.* v. *Kennett, R.* v. *Saunders,* L. R. 4 Q. B. 565; 10 B. & S. 545; 20 L. T. N.S. 656; 33 J. P. 824).

A baker was in the habit of selling loaves of bread at sixpence, varying the weight of the loaf with the price of corn. When he proposed to sell a three-and-a-half-pound loaf for sixpence his custom was to put into the oven four pounds of dough, but the loaf was not weighed after baking. Having sold, at sixpence each, six loaves, which varied in weight, but all but one weighed over three and a half pounds, he was convicted of selling bread otherwise than by weight, and the conviction was upheld (*Hill* v. *Browning,* L. R. 5 Q. B. 453; 22 L. T. N.S. 584; 34 J. P. 774).

The Aerated Bread Company made two kinds of bread, loaves baked in tins, which they called household bread, and which they sold by weight, and bread made in separate loaves, which were put separately in the oven so as to be baked crusty all over, which they called French or fancy bread, and this they did not sell by weight. The material of which the bread called fancy bread was made in no way differed from the ordinary loaves sold by bakers generally, and, except in the manner of baking it in separate loaves, it in no way resembled what was called French or fancy bread at the time of the passing of the Act, and was in fact only English bread baked so as to have an outside of crust. It was held that the loaves in question were not "French or fancy bread" within the exception of the Act (*Aerated Bread Company* v. *Gregg,* L. R. 8 Q. B. 355; 42 L. J. M. C. 117; 37 J. P. 388).

All bakers or sellers of bread, in the sale of bread shall use the avoirdupois weight of sixteen ounces to the pound, according to the standard in the Exchequer, and the several gradations of the same for any less quantity than a pound; and in case any such baker or seller of bread shall at any time use any other than the avoirdupois weight, and the usual gradations of the same, he, she, or they shall, for every such offence, forfeit and pay any sum not exceeding five pounds nor less than forty shillings, as the magistrate or magistrates, justice or justices, before whom such conviction shall take place, shall from time to time order and adjudge.

Bakers to use avoirdupois weight. 6 & 7 Will. IV. c. 37, s. 5.

Bakers to provide in their shops beams, scales and weights, &c., and to weigh bread, &c.
6 & 7 Will. IV. c. 37, s. 6.

Every baker or seller of bread shall cause to be fixed in some conspicuous part of his, her, or their shop, on or near the counter, a beam and scales with proper weights, or other sufficient balance, in order that all bread there sold may from time to time be weighed in the presence of the purchaser or purchasers thereof, except as aforesaid; and in case any such baker or seller of bread shall neglect to fix such beam and scales, or other sufficient balance, in manner aforesaid, or to provide and keep for use proper beam and scales and proper weights or balance, or shall have or use any incorrect or false beam or scales or balance, or any false weight not being of the weight it purports to be, according to the standard in the Exchequer, then and in every such case he, she, or they shall, for every such false beam and scales and balance, or false weight, forfeit and pay any sum not exceeding five pounds, which the magistrate or magistrates, justice or justices, before whom such offender or offenders shall be convicted, shall order and direct.

Bakers and sellers of bread, &c., delivering by cart, &c., to be provided with scales and weights, &c., for weighing bread.
6 & 7 Will. IV. c. 37, s. 7.

Every baker or seller of bread, and every journeyman, servant, or other person employed by such baker or seller of bread, who shall convey or carry out bread for sale in and from any cart or other carriage, shall be provided with, and shall constantly carry in such cart or other carriage, a correct beam and scales with proper weights, or other sufficient balance, in order that all bread sold by every such baker or seller of bread, or by his or her journeyman, servant, or other person, may from time to time be weighed in the presence of the purchaser or purchasers thereof, except as aforesaid; and in case any such baker or seller of bread, or his or her journeyman, servant, or other person, shall at any time carry out or deliver any bread, without being provided with such beam and scales, with proper weights, or other sufficient balance, or whose weights shall be deficient in their due weight according to the standard in the Exchequer, or shall at any time refuse to weigh any bread purchased of him, her, or them, or delivered by his, her, or their journeyman, servant, or other person, in the presence of the person or persons purchasing or receiving the same; then and in every such case every such baker or seller of bread shall, for every such offence, forfeit and pay any sum not exceeding five pounds, which the magistrate or magistrates, justice or justices, before whom such offender or offenders shall be convicted, shall order and direct.

Refuse to weigh in the presence of the purchaser.—A baker is liable to this penalty for refusing to weigh in the presence of the purchaser whether the bread be sold in a shop or from any cart or carriage, the offence not being confined to a sale of the latter description (*R.* v. *Kingsley*, 16 L. T. 408; 15 J. P. 65).

A baker carrying on business at a shop, under a course of dealing with a customer which had continued for two years, supplied bread to a customer at her house, always entering in his book the amount of bread left from time to time. He carried out this bread in a cart without being provided with weights and scales. Held that he was liable to be convicted under this section (*Robinson* v. *Cliff*, 45 L. J. M. C. 109; 40 J. P. 615).

The corresponding section in the Act of 1822 uses the term "cart or other carriage drawn by a horse, mule, or ass," the words "drawn by a horse, mule, or ass," having been omitted in the later Act of 1836. The effect of this omission is that in the Metropolitan area bakers selling bread or conveying bread for sale in a *barrow* need not be provided with scales under this section, but throughout the rest of England and Scotland the selling from "any cart or other carriage" would render the provision of weights and scales, &c., necessary.

In both of the Acts will be found the necessary legal provisions for the recovery and application of the penalties, appeal to quarter sessions, &c., &c.

APPENDIX.

1. THE WEIGHTS AND MEASURES ACT, 1878, 41 & 42 VICT. c. 49.

2. THE ANNOYANCE JURORS (WESTMINSTER) ACT, 1861, 24 & 25 VICT. c. 78.

3. FORMS:

   (i.) WARRANT FOR INSPECTOR.

   (ii.) CERTIFICATE OF STAMPING.

4. LIST OF OFFENCES.

5. ORDERS IN COUNCIL.

WEIGHTS AND MEASURES ACT, 1878.
[41 & 42 VICT. CH. 49.]

ARRANGEMENT OF SECTIONS.

Preliminary.

Section.
1. Short title.
2. Commencement.

I.—LAW OF WEIGHTS AND MEASURES.

Uniformity of Weights and Measures.

3. Uniformity of weights and measures.

Standards of Measure and Weight.

4. Imperial standards of measure and weight.
5. Parliamentary copies of imperial standards.
6. Restoration of imperial standards.
7. Restoration of parliamentary copies.
8. Secondary (Board of Trade) standards of measure and weight.
9. Local standards of measure and weight.

Imperial Measures of Length.

10. Imperial standard yard.
11. Linear measures derived from imperial standard yard.
12. Superficial measures derived from the imperial standard yard.

Imperial Measures of Weight and Capacity.

13. Imperial standard pound.
14. Imperial weights derived from imperial standard pound.
15. Imperial measures of capacity.
16. Measure of capacity for goods formerly sold by heaped measure.
17. Measure of capacity when used to be stricken or filled up.

Metric Equivalents of Imperial Weights and Measures.

Section.
18. Equivalents of metric weights and measures in terms of imperial weights and measures.

Use of Imperial Weights and Measures.

19. Trade contracts, sales, dealings, &c. to be in terms of imperial weights or measures.
20. Sale by avoirdupois weight, with exceptions.
21. Exception for contract, &c. in metric weights and measures.
22. Exception for sale of article in vessel not represented as being of imperial or local measure.
23. Penalty on price lists, &c. denoting greater or less weight or measure than the same denomination of imperial weight or measure.
24. Penalty on use or possession of unauthorised weight or measure.

Unjust Weights and Measures.

25. Penalty on use or possession of unjust measures, weights, balances, or weighing machines.
26. Penalty for fraud in use of weight, measure, balance, &c.
27. Penalty on sale of false weight, measure, balance, &c.

Stamping and Verification of Weights and Measures.

28. Stamping of weights and measures with denomination.
29. Stamping of verification on measures and weights.
30. Lead or pewter weights.
31. Stamping of verification on weights for coin.
32. Forgery, &c. of stamps on measures or weights.

II.—ADMINISTRATION.

(a.) *Central.*

Board of Trade.

33. Powers and duties of Board of Trade as to standards of weights and measures, &c.

Custody and Verification of Standards and Copies.

34. Custody of imperial and Board of Trade standards to remain with Board of Trade.
35. Custody and periodical verification of parliamentary copies of imperial standards.
36. Periodical verification of Board of Trade standards.
37. Verification by Board of Trade of local standards.
38. Power of Board of Trade to verify metric weights and measures.
39. Verification and stamping of coin weights.

Appendix.

(b.) *Local Administration.*

Local Standards.

Section.
40. Provision of local standards by local authority.
41. Periodical verification of local standards.
42. Production of local standards.

Local Verification and Inspection of Weights and Measures.

43. Appointment of inspectors of weights and measures.
44. Verification and stamping by inspectors of weights and measures.
45. Validity of weights and measures stamped throughout the United Kingdom.
46. Power to stamp measures made partly of metal and partly of glass.
47. Fees for comparison and stamping.
48. Power to inspect measures, weights, scales, &c. and to enter shops, &c. for that purpose.
49. Penalty on inspector for misconduct.

Local Authorities.

50. Local authorities and local rate.
51. Expenses of local authority.
52. Power of local authorities to combine for purposes of Act.
53. Power to local authority to make bye-laws as to local verification, &c.
54. Appointment of inspectors in towns and other places.
55. Power of vestry, &c. in Metropolis to put an end to appointment of inspectors of weights and measures under Local Act.

Legal Proceedings.

56. Prosecution of offences and recovery of fines.
57. Provisions as to summary proceedings.
58. Limitation as to conviction for second offences.
59. Evidence as to possession.
60. Appeal from conviction.
61. Provision as to action against person acting in execution of Act.

III.—MISCELLANEOUS.

62. Continuance of inquisition recorded for ascertaining rents and tolls payable.
63. Orders in Council.
64. Effect of schedules.
65. Construction of Acts referring to repealed enactments.

Savings and Definitions.

Section.
66. Saving as to models of gas holders under 22 & 23 Vict. c. 66.
67. Saving as to rights of the Founders Company.
68. Saving as to London.
69. Act not to abridge the power of the leet jury, &c.
70. Definitions.

IV.—Application of Act to Scotland.

71. Application of imperial weights and measures to tolls, &c.
72. Recovery and application of penalties.
73. Appeal.
74. Definitions as regards Scotland.
75. Power of sheriff.

V.—Application of Act to Ireland.

76. Contracts to be made by denominations of imperial weight, otherwise to be void.
77. Mode of weighing. Deductions prohibited.
78. Providing of local standards and sub-standards.
79. Inquiry by judge of assize and chairman of quarter sessions as to provision of local standards and sub-standards.
80. Expenses of ex-officio inspectors.
81. Ex-officio inspectors of weights and measures.
82. Custody and use of local standards.
83. Custody and periodical verification of local sub-standards.
84. Recovery of fines, &c.
85. Definitions.

VI.—Repeal.

86. Repeal.

SCHEDULES.

CHAPTER 49.

An Act to consolidate the Law relating to Weights and Measures. [8th August 1878.]

BE it enacted by the Queen's most Excellent Majesty, by and with the advice and consent of the Lords Spiritual and Temporal, and Commons, in this present Parliament assembled, and by the authority of the same, as follows:

Preliminary.

1. This Act may be cited as the Weights and Measures Act, 1878. — *Short title.*

2. This Act shall not come into operation until the first day of January one thousand eight hundred and seventy-nine, which day is herein-after referred to as the commencement of this Act. — *Commencement.*

I.—LAW OF WEIGHTS AND MEASURES.

Uniformity of Weights and Measures.

3. The same weights and measures shall be used throughout the United Kingdom. — *Uniformity of weights and measures.*

Standards of Measure and Weight.

4. The bronze bar and the platinum weight, more particularly described in the first part of the First Schedule to this Act, and at the passing of this Act deposited in the Standards Department of the Board of Trade in the custody of the Warden of the Standards, shall continue to be the imperial standards of measure and weight, and the said bronze bar shall continue to be the imperial standard for determining the imperial standard yard for the United Kingdom, and the said platinum weight shall continue to be the imperial standard for determining the imperial standard pound for the United Kingdom. — *Imperial standards of measure and weight.*

5. The four copies of the imperial standards of measure and weight, described in the second part of the First Schedule to this Act, and deposited as therein mentioned, shall be deemed to be parliamentary copies of the said imperial standards. — *Parliamentary copies of imperial standards.*

The Board of Trade shall as soon as may be after the commencement of this Act cause an accurate copy of the imperial standard of measure and an accurate copy of the imperial standard of weight to be made of the same form and material as the said standards, and it shall be lawful for Her Majesty in Council, on the representation of the Board of Trade, to approve the copies so made, and the copies when so approved shall be of the same effect as the said parliamentary copies, and are in this Act included under the name parliamentary copies of the imperial standards of measure and weight.

Restoration of imperial standards.

6. If at any time either of the imperial standards of measure and weight is lost or in any manner destroyed, defaced, or otherwise injured, the Board of Trade may cause the same to be restored by reference to or adoption of any of the parliamentary copies of that standard, or of such of them as may remain available for that purpose.

Restoration of parliamentary copies.

7. If at any time any of the parliamentary copies of either of the imperial standards is lost or in any manner destroyed, defaced, or otherwise injured, the Board of Trade may cause the same to be restored by reference either to the corresponding imperial standard, or to one of the other parliamentary copies of that standard.

Secondary (Board of Trade) standards of measure and weight.

8. The secondary standards of measure and weight which, having been derived from the imperial standards, are at the commencement of this Act in use under the direction of the Board of Trade, and are mentioned in the Second Schedule to this Act, and no others (save as herein-after mentioned), shall be secondary standards of measure and weight, and shall be called Board of Trade standards.

If at any time any of such standards is lost or in any manner destroyed, defaced, or otherwise injured, the Board of Trade may cause the same to be restored by reference either to one of the imperial standards or to one of the parliamentary copies of those standards.

The Board of Trade shall from time to time cause such new denominations of standards, being either equivalent to or multiples or aliquot parts of the imperial weights and measures ascertained by this Act, or being equivalent to or multiples of each coin of the realm for the time being, as appear to them to be required, in addition to those mentioned in the Second Schedule to this Act, to be made and duly verified, and those new denominations of standards when approved by Her Majesty in Council shall be Board of Trade standards in like manner as if they were mentioned in the said schedule.

It shall be lawful for Her Majesty by Order in Council to declare that a Board of Trade standard for the time being of any denomination, whether mentioned in the said schedule or approved by Order in Council, shall cease to be such a standard.

Such standards of the Board of Trade as are equivalent to or multiples of any coin of the realm for the time being shall be standard weights for determining the justness of the weight of and for weighing such coin.

Local standards of measure and weight.

9. The standards of measure and weight which are at the commencement of this Act legally in use by inspectors of weights and measures for the purpose of verification or inspection, and all copies of the Board of Trade standards which after the commencement of this Act are compared with those standards and verified by the Board of Trade for the purpose of being used by inspectors of

Appendix.

weights and measures under this Act as standards for the verification or inspection of weights and measures, shall be called local standards.

Imperial Measures of Length.

10. The straight line or distance between the centres of the two gold plugs or pins (as mentioned in the First Schedule to this Act) in the bronze bar by this Act declared to be the imperial standard for determining the imperial standard yard measured when the bar is at the temperature of sixty-two degrees of Fahrenheit's thermometer, and when it is supported on bronze rollers placed under it in such manner as best to avoid flexure of the bar, and to facilitate its free expansion and contraction from variations of temperature, shall be the legal standard measure of length, and shall be called the imperial standard yard, and shall be the only unit or standard measure of extension from which all other measures of extension, whether linear, superficial or solid, shall be ascertained. *Imperial standard yard.*

11. One third part of the imperial standard yard shall be a foot, and the twelfth part of such foot shall be an inch, and the rod, pole, or perch in length shall contain five such yards and a half, and the chain shall contain twenty-two such yards, the furlong two hundred and twenty such yards, and the mile one thousand seven hundred and sixty such yards. *Linear measures derived from imperial standard yard.*

12. The rood of land shall contain one thousand two hundred and ten square yards according to the imperial standard yard, and the acre of land shall contain four thousand eight hundred and forty such square yards, being one hundred and sixty square rods, poles, or perches. *Superficial measures derived from the imperial standard yard.*

Imperial Measures of Weight and Capacity.

13. The weight in vacuo of the platinum weight (mentioned in the First Schedule to this Act), and by this Act declared to be the imperial standard for determining the imperial standard pound, shall be the legal standard measure of weight, and of measure having reference to weight, and shall be called the imperial standard pound, and shall be the only unit or standard measure of weight from which all other weights and all measures having reference to weight shall be ascertained. *Imperial standard pound.*

14. One sixteenth part of the imperial standard pound shall be an ounce, and one sixteenth part of such ounce shall be a dram, and one seven-thousandth part of the imperial standard pound shall be a grain. *Imperial weights derived from imperial standard pound.*

A stone shall consist of fourteen imperial standard pounds, and a hundredweight shall consist of eight such stones, and a ton shall consist of twenty such hundredweights.

Four hundred and eighty grains shall be an ounce troy.

All the foregoing weights except the ounce troy shall be deemed to be avoirdupois weights.

15. The unit or standard measure of capacity from which all other measures of capacity, as well for liquids as for dry goods, shall be derived, shall be the gallon containing ten imperial stan- *Imperial measures of capacity.*

dard pounds weight of distilled water weighed in air against brass weights, with the water and the air at the temperature of sixty-two degrees of Fahrenheit's thermometer, and with the barometer at thirty inches.

The quart shall be one fourth part of the gallon, and the pint shall be one eighth part of the gallon.

Two gallons shall be a peck, and eight gallons shall be a bushel, and eight such bushels shall be a quarter, and thirty-six such bushels shall be a chaldron.

Measure of capacity for goods formerly sold by heaped measure.
5 & 6 Will. IV. c. 63.

16. A bushel for the sale of any of the following articles, namely, lime, fish, potatoes, fruit, or any other goods and things which before (the passing of the Weights and Measures Act, 1835, that is to say) the ninth day of September one thousand eight hundred and thirty-five, were commonly sold by heaped measure, shall be a hollow cylinder having a plane base, the internal diameter of which shall be double the internal depth, and every measure used for the sale of any of the above-mentioned articles which is a multiple of a bushel, or is a half bushel or a peck, shall be made of the same shape and proportion as the above-mentioned bushel.

Measure of capacity when used to be stricken or filled up.

17. In using an imperial measure of capacity, the same shall not be heaped, but either shall be stricken with a round stick or roller, straight and of the same diameter from end to end, or if the article sold cannot from its size or shape be conveniently stricken, shall be filled in all parts as nearly to the level of the brim as the size and shape of the article will admit.

Metric Equivalents of Imperial Weights and Measures.

Equivalents of metric weights and measures in terms of imperial weights and measures.

18. The table in the third schedule to this Act shall be deemed to set forth the equivalents of imperial weights and measures and of the weights and measures therein expressed in terms of the metric system, and such table may be lawfully used for computing and expressing, in weights and measures, weights and measures of the metric system.

Use of Imperial Weights and Measures.

Trade contracts, sales, dealings, &c., to be in terms of imperial weights or measures.

19. Every contract, bargain, sale, or dealing, made or had in the United Kingdom for any work, goods, wares, or merchandise, or other thing which has been or is to be done, sold, delivered, carried, or agreed for by weight or measure, shall be deemed to be made and had according to one of the imperial weights or measures ascertained by this Act, or to some multiple or part thereof, and if not so made or had shall be void; and all tolls and duties charged or collected according to weight or measure shall be charged and collected according to one of the imperial weights or measures ascertained by this Act, or to some multiple or part thereof.

Such contract, bargain, sale, dealing, and collection of tolls and duties as is in this section mentioned is in this Act referred to under the term "trade."

No local or customary measures, nor the use of the heaped measure, shall be lawful.

Any person who sells by any denomination of weight or measure other than one of the imperial weights or measures, or some multiple or part thereof, shall be liable to a fine not exceeding forty shillings for every such sale.

20. All articles sold by weight shall be sold by avoirdupois weight; except that— *Sale by avoirdupois weight, with exceptions.*

1. Gold and silver, and articles made thereof, including gold and silver thread, lace, or fringe, also platinum, diamonds, and other precious metals or stones, may be sold by the ounce troy or by any decimal parts of such ounce; and all contracts, bargains, sales, and dealings in relation thereto shall be deemed to be made and had by such weight, and when so made or had shall be valid; and
2. Drugs, when sold by retail, may be sold by apothecaries weight.

Every person who acts in contravention of this section shall be liable to a fine not exceeding five pounds.

21. A contract or dealing shall not be invalid or open to objection on the ground that the weights or measures expressed or referred to therein are weights or measures of the metric system, or on the ground that decimal subdivisions of imperial weights and measures, whether metric or otherwise, are used in such contract or dealing. *Exception for contract, &c., in metric weights and measures.*

22. Nothing in this Act shall prevent the sale, or subject a person to a fine under this Act for the sale, of an article in any vessel, where such vessel is not represented as containing any amount of imperial measure, nor subject a person to a fine under this Act for the possession of a vessel where it is shown that such vessel is not used nor intended for use as a measure. *Exception for sale of article in vessel not represented as being of imperial or local measure.*

23. Any person who prints, and any clerk of a market or other person who makes, any return, price list, price current, or any journal or other paper containing price list or price current, in which the denomination of weights and measures quoted or referred to denotes or implies a greater or less weight or measure than is denoted or implied by the same denomination of the imperial weights and measures under this Act, shall be liable to a fine not exceeding ten shill'ngs for every copy of every such return, price list, price current, journal, or other paper which he publishes. *Penalty on price lists, &c., denoting greater or less weight or measure than the same denomination of imperial weight or measure.*

24. Every person who uses or has in his possession for use for trade a weight or measure which is not of the denomination of some Board of Trade standard, shall be liable to a fine not exceeding five pounds, or in the case of a second offence ten pounds, and the weight or measure shall be liable to be forfeited. *Penalty on use or possession of unauthorised weight or measure.*

Unjust Weights and Measures.

25. Every person who uses or has in his possession for use for trade any weight, measure, scale, balance, steelyard, or weighing *Penalty on use or possession of*

unjust measures, weights, balances, or weighing machines.
Penalty for fraud in use of weight, measure, balance, &c.

machine which is false or unjust, shall be liable to a fine not exceeding five pounds, or in the case of a second offence ten pounds, and any contract, bargain, sale, or dealing made by the same shall be void, and the weight, measure, scale, balance, or steelyard shall be liable to be forfeited.

26. Where any fraud is wilfully committed in the using of any weight, measure, scale, balance, steelyard, or weighing machine, the person committing such fraud, and every person party to the fraud, shall be liable to a fine not exceeding five pounds, or in the case of a second offence ten pounds, and the weight, measure, scale, balance, or steelyard, shall be liable to be forfeited.

Penalty on sale of false weight, measure, balance, &c.

27. A person shall not wilfully or knowingly make or sell, or cause to be made or sold, any false or unjust weight, measure, scale, balance, steelyard, or weighing machine.

Every person who acts in contravention of this section shall be liable to a fine not exceeding ten pounds, or in the case of a second offence fifty pounds.

Stamping and Verification of Weights and Measures.

Stamping of weights and measures with denomination.

28. Every weight, except where the small size of the weight renders it impracticable, shall have the denomination of such weight stamped on the top or side thereof in legible figures and letters.

Every measure of capacity shall have the denomination thereof stamped on the outside of such measure in legible figures and letters.

A weight or measure not in conformity with this section shall not be stamped with such stamp of verification under this Act as is herein-after mentioned.

Stamping of verification on measures and weights.

29. Every measure and weight whatsoever used for trade shall be verified and stamped by an inspector with a stamp of verification under this Act.

Every person who uses or has in his possession for use for trade any measure or weight not stamped as required by this section, shall be liable to a fine not exceeding five pounds, or in the case of a second offence ten pounds, and shall be liable to forfeit the said measure or weight, and any contract, bargain, sale, or dealing made by such measure or weight shall be void.

Lead or pewter weights.

30. A weight made of lead or pewter, or of any mixture thereof, shall not be stamped with a stamp of verification or used for trade, unless it be wholly and substantially cased with brass, copper, or iron, and legibly stamped or marked " cased " :

Provided that nothing in this section shall prevent the insertion into a weight of such a plug of lead or pewter as is *bonâ fide* necessary for the purpose of adjusting it and of affixing thereon the stamp of verification.

Stamping of verification on weights for coin.

A person guilty of any offence against or disobedience to the provisions of this section, shall be liable to a penalty not exceeding five pounds, or in case of a second offence ten pounds.

Appendix.

31. Every coin weight, not less in weight than the weight of the lightest coin for the time being current, shall be verified and stamped by the Board of Trade with a mark of verification under this Act, and otherwise shall not be deemed a just weight for determining the weight of gold and silver coin of the realm.

Every person who uses any weight declared by this section not to be a just weight shall be liable to a fine not exceeding fifty pounds.

32. If any person forges or counterfeits any stamp used for the stamping under this Act of any measure or weight, or used before the commencement of this Act for the stamping of any measure or weight, under any enactment repealed by this Act, or wilfully increases or diminishes a weight so stamped, he shall be liable to a fine not exceeding fifty pounds. *Forgery, &c., of stamps on measures or weights.*

Any person who knowingly uses, sells, utters, disposes of, or exposes for sale any measure or weight with such forged or counterfeit stamp thereon, or a weight so increased or diminished, shall be liable to a fine not exceeding ten pounds.

All measures and weights with any such forged or counterfeit stamp shall be forfeited.

II.—Administration.

(a.) Central.

Board of Trade.

33. The Board of Trade shall have all such powers and perform all such duties relative to standards of measure and weight, and to weights and measures, as are by any Act or otherwise vested in or imposed on the Treasury, or the Comptroller-General of the Exchequer, or the Warden of the Standards; and all things done by the Board of Trade, or any of their officers, or at their office, in relation to standards of weights and measures in pursuance of this Act shall be valid, and have the like effect and consequences, as if the same had been done by the Treasury, or by the Comptroller-General or other officer of the Exchequer, or by the Warden of the Standards, or at the office of the Exchequer. *Powers and duties of Board of Trade as to standards of weights and measures, &c.*

It shall be the duty of the Board of Trade to conduct all such comparisons, verifications, and other operations with reference to standards of measure and weight, in aid of scientific researches or otherwise, as the Board of Trade from time to time thinks expedient, and to make from time to time a report to Parliament on their proceedings and business under this Act.

Custody and Verification of Standards and Copies.

34. The imperial standards of measure and weight, the Board of Trade standards of measure and weight, and all balances, apparatus, books, documents, and things used in connection therewith or relating thereto, deposited at the passing of this Act in the Standards Department, or in any other office of the Board of Trade, shall remain and be in the custody of the Board of Trade. *Custody of imperial and Board of Trade standards to remain with Board of Trade.*

Custody and periodical verification of parliamentary copies of imperial standards.	**35.** The parliamentary copies of the imperial standards of measure and weight mentioned in part two of the First Schedule to this Act shall continue to be deposited as therein mentioned.

The copies of the imperial standards of measure and weight made in pursuance of this Act, when approved by Her Majesty in Council, shall be deposited at some office of the Board of Trade, and be in the custody of the Board of Trade.

The Board of Trade shall cause the parliamentary copies of the imperial standards of measure and weight, except the copy immured in the new palace at Westminster, to be compared once in every ten years with each other, and once in every twenty years with the imperial standards of measure and weight.

Periodical verification of Board of Trade standards.	**36.** Once at least in every five years the Board of Trade shall cause the Board of Trade standards for the time being to be compared with the parliamentary copies of the imperial standards of measure and weight made and approved in pursuance of this Act and with each other, and to be adjusted or renewed, if requisite.
Verification by Board of Trade of local standards.	**37.** The Board of Trade shall cause to be compared with the Board of Trade standards, and verified at such place as the Board of Trade in each case direct, all copies of any of those standards which are submitted for the purpose by any local authority, and have been used or are intended to be used as local standards, and if they find the same fit for the purpose of being used by inspectors of weights and measures under this Act as standards for the verification and inspection of weights and measures, shall cause them to be stamped as verified or re-verified in such manner as to show the date of such verification or re-verification, and every such verification shall be evidenced by an indenture, and every such re-verification shall be evidenced by an indorsement upon the original indenture of verification, or by a new indenture of verification.

Any such indenture or indorsement, if purporting to be signed (either before or after the passing of this Act) by an officer of the Board of Trade, shall be evidence of the verification or re-verification of the weights and measures therein referred to.

Any such indenture or indorsement shall not be liable to stamp duty, nor shall any fee be payable on the verification or re-verification of any local standard.

An account shall be kept by the Board of Trade of all local standards verified or re-verified.

Power of Board of Trade to verify metric weights and measures.	**38.** Whereas the Board of Trade have obtained accurate copies of the metric standards mentioned in part two of the Third Schedule to this Act, and it is expedient to make the provision hereinafter mentioned for the verification of metric weights and measures, be it therefore enacted as follows:

The Board of Trade may, if they think fit, cause to be compared with the metric standards in their custody and verified all metric weights and measures which are submitted to them for the purpose, and are of such shape and construction as may be from time to time in that behalf directed by the Board of Trade, and which the Board of Trade are satisfied are intended to be used for the purpose of science or of manufacture, or for any lawful purpose not being for the purpose of trade within the meaning of this Act.

39. The Board of Trade, on payment of such fee, not exceeding five shillings. as they from time to time prescribe, shall cause all coin weights required by this Act to be verified, to be compared with the standard weights for weighing coin, and, if found to be just, stamped with a mark approved of by the Board, and notified in the London Gazette. *Verification and stamping of coin weights.*

All fees under this section shall be paid into the Exchequer.

(b.) *Local Administration.*

Local Standards.

40. The local authority (mentioned in the Fourth Schedule to this Act) of every county and borough from time to time shall provide such local standards of measure and weight as they deem requisite for the purpose of the comparison by way of verification for inspection, in accordance with this Act, of all weights and measures in use in their county or borough, and shall fix the places at which such standards are to be deposited. *Provision of local standards by local authority.*

The said local authority shall also provide from time to time proper means for verifying weights and measures by comparison with the local standards of such authority, and for stamping the weights and measures so verified.

41. A local standard of weight shall not be deemed legal nor be used for the purposes of this Act unless it has been verified or re-verified within five years before the time at which it is used. *Periodical verification of local standards.*

A local standard of measure shall not be deemed legal nor be used for the purposes of this Act unless it has been verified or re-verified within ten years before the time at which it is used.

A local standard of weight or measure which has become defective in consequence of any wear or accident, or has been mended, shall not be legal nor be used for the purpose of this Act until it has been re-verified by the Board of Trade.

A local standard may, save as aforesaid, be re-verified, for the purpose of this section, by such local comparison thereof as is hereinafter mentioned, if on that local comparison it is found correct, but otherwise shall be, and in any case may be, re-verified by the Board of Trade.

A local comparison of a local standard shall be made by an inspector of weights and measures for the county or borough in which such standard is used comparing the same, in the presence of a justice of the peace, with some other local standard which has been verified or re-verified by the Board of Trade, in the case of a weight within the previous five years, and in the case of a measure within the previous ten years.

Upon a local comparison where the local standard is found correct the justice shall sign an indorsement upon the indenture of verification of that standard, stating such local comparison and verification, and the error, if any, found thereon, and the indorsement so signed shall be transmitted to the Board of Trade to be recorded in the account of the verification of local standards. The indorsement when so recorded shall be evidence of the local comparison and verification, and a statement of the record thereof, if purporting to be signed by an officer of the Board of Trade, shall be evidence of the same having been so recorded.

It shall be lawful for her Majesty from time to time, by Order in Council, to define the amount of error to be tolerated in local standards when verified or re-verified by the Board of Trade, or when re-verified by such a local comparison as is authorised by this section.

Production of local standards.

42. The local standards shall be produced by the person having the custody thereof, upon reasonable notice, at such reasonable time and place within the county, borough or place for which the same have been provided, as any person by writing under his hand requires, upon payment by the person requiring such production of the reasonable charges of producing the same.

Local Verification and Inspection of Weights and Measures.

Appointment of inspectors of weights and measures.

43. Every local authority shall from time to time appoint a sufficient number of inspectors of weights and measures for safely keeping the local standards provided by such authority, and for the discharge of the other duties of inspectors under this Act; and where they appoint more than one such inspector, shall allot to each inspector (subject to any arrangement made for a chief inspector or inspectors) a separate district, to be distinguished by some name, number, or mark; and the local authority may suspend or dismiss any inspector appointed by them or appoint additional inspectors, as occasion may require, and shall assign reasonable remuneration to each inspector for his duties.

A local authority may, if they think fit, appoint different persons to be inspectors for verification and for inspection respectively of weights and measures under this Act.

A maker or seller of weights or measures, or a person employed in the making or selling thereof, shall not be an inspector of weights and measures under this Act.

An inspector of weights and measures shall forthwith on his appointment enter into a recognizance to the Crown (to be sued for in any court of record) in the sum of two hundred pounds for the due performance of the duties of his office, and for the due payment at the times fixed by the local authority appointing him, of all fees received by him under this Act, and for the safety of the local standards and the stamps and appliances for verification committed to his charge, and for their due surrender immediately on his removal or other cessation from office to the person appointed by the local authority to receive them.

Verification and stamping by inspectors of weights and measures.

44. The local authority shall from time to time fix the times and places within their jurisdiction at which each inspector appointed by them is to attend for the purpose of the verification of weights and measures; and the inspector shall attend, with the local

Appendix.

standards in his custody, at each time and place fixed, and shall examine every measure or weight which is of the same denomination as one of such standards, and is brought to him for the purpose of verification, and compare the same with that standard, and if he find the same correct shall stamp it with a stamp of verification in such manner as best to prevent fraud ; and in the case of a measure or of a weight of a quarter of a pound or upwards, shall further stamp thereon a name, number, or mark distinguishing the district for which he acts.

He shall also enter in a book kept by him minutes of every such verification, and give, if required, a certificate under his hand of every such stamping.

An inspector appointed by the local authority for a county may enter a place within the district of an inspector appointed by any other local authority, and there verify and stamp the weights and measures of any person residing within his own district, but if he knowingly stamp a weight or measure of any person residing in the district of an inspector legally appointed by another local authority, he shall be liable to a fine not exceeding twenty shillings for every weight or measure which he so stamps.

45. A weight or measure duly stamped by an inspector under this Act shall be a legal weight or measure throughout the United Kingdom, unless found to be false or unjust, and shall not be liable to be re-stamped because used in any place other than that in which it was originally stamped. Validity of weights and measures stamped [throughout the United Kingdom.

46. Where a measure for liquids is constructed with a small window or transparent part through which the contents, whether to the brim or to any other index thereof, may be seen without impediment, such measure may be verified and stamped by inspectors under this Act, although such measure is made partly of metal and partly of glass or other transparent medium, and that whether such measure corresponds exactly to a Board of Trade standard, or whether it exceeds such standard, but has the capacity of such standard indicated by a level line drawn through the centre of the window or transparent part. Power to stamp measures made partly of metal and partly of glass.

47. An inspector under this Act may take in respect of the verification and stamping of weights and measures such fees not exceeding those specified in the Fifth Schedule to this Act as the authority appointing him from time to time fix, and shall at such times not less often than once a quarter as the said authority direct, account for and pay over to the treasurer of the local rate or such person as the said authority direct all fees taken by him. Fees for comparison and stamping.

Where the Board of Trade, upon the application of any local authority from time to time represent to Her Majesty that it would be expedient to alter the fees taken by the inspectors of such authority under this Act (whether specified in the said schedule or in any order previously made under this section) or, for the purpose of adapting those fees to the local standards provided by such authority, to add to the said fees, it shall be lawful for Her Majesty by Order in Council from time to time to alter or add to the said fees.

Appendix.

Power to inspect measures, weights, scales, &c., and to enter shops, &c., for that purpose.

48. Every inspector under this Act authorised in writing under the hand of a justice of the peace, also every justice of the peace, may at all reasonable times inspect all weights, measures, scales, balances, steelyards and weighing machines within his jurisdiction which are used or in the possession of any person or on any premises for use for trade, and may compare every such weight and measure with some local standard, and may seize and detain any weight, measure, scale, balance or steelyard, which is liable to be forfeited in pursuance of this Act, and may for the purpose of such inspection enter any place, whether a building or in the open air, whether open or enclosed, where he has reasonable cause to believe that there is any weight, measure, scale, balance, steelyard or weighing machine which he is authorised by this Act to inspect.

Any person who neglects or refuses to produce for such inspection all weights, measures, scales, balances, steelyards and weighing machines in his possession or on his premises, or refuses to permit the justice or inspector to examine the same or any of them, or obstructs the entry of the justice or inspector under this section, or otherwise obstructs or hinders a justice or inspector acting under this section, shall be liable to a fine not exceeding five, or, in the case of a second offence, ten pounds.

Penalty on inspector for misconduct.

49. If an inspector under this Act stamps a weight or measure in contravention of any provision of this Act, or without duly verifying the same by comparison with a local standard, or is guilty of a breach of any duty imposed on him by this Act, or otherwise misconducts himself in the execution of his office, he shall be liable to a fine not exceeding five pounds for each offence.

Local Authorities.

Local authorities and local rate.

50. For the purposes of this Act "the local authority" and "the local rate" shall mean in each of the different areas mentioned in the first column of the Fourth Schedule to this Act the authority and the rate or fund mentioned in that schedule in connection with that area:

Provided that in England the council of a borough which has not a separate court of quarter sessions shall not, unless they so resolve, be the local authority for the purposes of this Act, and if they so resolve and provide local standards and appoint inspectors after the commencement of this Act, they shall forthwith give notice of such resolution and appointment, under the corporate seal of the borough, to the clerk of the peace of the county in which the borough is situate, and after the expiration of one month from the day on which that notice of the said appointment is given the powers of inspectors of weights and measures appointed by the justices of the county shall, as to such borough and the weights and measures of persons residing therein, cease; but until such notice is given the borough shall be deemed to form part of the said county in like manner as if the same were not a borough.

Where at the commencement of this Act legal local standards are provided and inspectors are appointed by the council of a borough not having a separate court of quarter sessions, that council shall continue to be the local authority until they otherwise resolve.

51. The expense of providing and re-verifying local standards, the salaries of the inspectors, and all other expenses incurred by the local authority under this Act shall be paid out of the local rate. — *Expenses of local authority.*

The treasurer of the county in which a borough in England having a separate court of quarter sessions is situate shall exclude from the account kept by him of all sums expended out of the county rate to which the borough is liable to contribute all sums expended in pursuance of this Act.

52. Any two or more local authorities may combine, as regards either the whole or any part of the areas within their jurisdiction, for all or for any of the purposes of this Act, upon such terms and in such manner as may be from time to time mutually agreed upon. — *Power of local authorities to combine for purposes of Act.*

An inspector appointed in pursuance of an agreement for such combination shall, subject to the terms of his appointment, have the same authority jurisdiction and duties as if he had been appointed by each of the authorities who are parties to such agreement.

53. Any local authority from time to time, with the approval of the Board of Trade, may make, and when made, revoke, alter, and add to, byelaws for regulating the comparison with the local standards of such authority, and the verification and stamping of weights and measures in use in their county or borough, and for regulating the local comparison of the local standards of such authority, and generally for regulating the duties under this Act of the inspectors appointed by the local authority or of any of those inspectors. Such byelaws may impose fines not exceeding twenty shillings for the breach of any byelaw, to be recovered on summary conviction. The Board of Trade before approving any such byelaws shall cause them to be published in such manner as they think sufficient for giving notice thereof to all persons interested. — *Power to local authority to make byelaws as to local verification, &c.*

54. Where a town or other place has been or may hereafter be authorised under any Act, whether local or otherwise, to appoint inspectors or examiners of weights and measures, or where any other place has been or may hereafter be, by charter, Act of Parliament or otherwise, possessed of legal jurisdiction, and such town or place is for the time being provided with legal local standards, the magistrates of such town or place, or other persons authorised as aforesaid, may appoint inspectors of weights and measures within the limits of their jurisdiction, and suspend and dismiss such inspectors, and such inspectors shall within such limits exclusively have the same power and discharge the same duties as inspectors of weights and measures appointed under this Act by the local authorities for the county, and shall pay over and account for the fees received by them under this Act, to such persons as may be duly authorised by the magistrates or other persons appointing them. — *Appointment of inspectors in towns and other places.*

55. Where in any place in the Metropolis—that is to say, in the parishes and places in which the Metropolitan Board of Works have power to levy the consolidated rate—any vestry, commissioners — *Power of vestry, &c., in Metropolis to*

put an end to appointment of inspectors of weights and measures under Local Act. or other body have any duties or powers, under any Local Act, charter or otherwise, in relation to the appointment of inspectors or examiners of weights and measures, such vestry, commissioners or body may, at a meeting specially convened for the purpose, of which not less than fourteen days notice has been given, resolve that it is expedient that their said duties and powers should cease in such place.

The clerk or other like officer of such vestry, commissioners or body shall give notice of such resolution to the clerk of the peace for the county in which such place is situate, and the clerk of the peace shall lay such notice before the next practicable court of quarter sessions for the county, and after the receipt of such notice by the court of quarter sessions the appointment, and all powers of appointment, of any inspector or examiner appointed under such Local Act, charter or otherwise, shall cease in the said place, without prejudice to any proceedings then pending for penalties or otherwise.

Legal Proceedings.

Prosecution of offences and recovery of fines.
56. All offences under this Act may be prosecuted, and all fines and forfeitures under this Act may be recovered on summary conviction before a court of summary jurisdiction in manner provided by the Summary Jurisdiction Act.

The court when hearing and determining an information or complaint under this Act shall be constituted either of two or more justices of the peace in petty sessions sitting at a place appointed for holding petty sessions, or of some magistrate or officer sitting alone or with others at some court or other place appointed for the administration of justice and for the time being empowered by law to do alone any act authorised to be done by more than one justice of the peace.

Provisions as to summary proceedings.
57. The following enactments shall apply to proceedings under this Act before a court of summary jurisdiction; (that is to say,)
1. The description of any offence in the words of this Act, or in similar words, shall be sufficient in law; and
2. Any exception, exemption, proviso, excuse, or qualification, whether it does or does not accompany in the same section the description of the offence, may be proved by the defendant but need not be specified or negatived in the information or complaint, and, if so specified or negatived, no proof in relation to the matter so specified or negatived shall be required on the part of the informant or complainant; and
3. A warrant of commitment shall not be held void by reason of any defect therein, if it be therein alleged that the offender has been convicted, and there is a good and valid conviction to sustain the same.
4. Such portion of any fine under this Act, not exceeding a moiety, as the court of summary jurisdiction before whom a person is convicted think fit to direct, may, if the court in their discretion so order, be paid to the informer.
5. All weights, measures, scales, balances and steelyards forfeited

under this Act shall be broken up, and the materials thereof may be sold or otherwise disposed of as a court of summary jurisdiction direct, and the proceeds of such sale shall be applied in like manner as fines under this Act.

58. A person shall not be liable to any increased penalty for a second offence under any section of this Act, unless that offence was committed after a conviction within five years previously for an offence under the same section. *Limitation as to conviction for second offences.*

59. Where any weight, measure, scale, balance, steelyard or weighing machine is found in the possession of any person carrying on trade within the meaning of this Act, or on the premises of any person which, whether a building or in the open air, whether open or enclosed, are used for trade within the meaning of this Act, such person shall be deemed for the purposes of this Act, until the contrary is proved, to have such weight, measure, scale, balance, steelyard or weighing machine in his possession for use for trade. *Evidence as to possession.*

60. Any person who feels himself aggrieved by a conviction or order of a court of summary jurisdiction under this Act may appeal therefrom, subject in England to the conditions following; that is to say, *Appeal from conviction.*

1. The appeal shall be made to the next practicable court of general or quarter sessions having jurisdiction in the county or place in which the decision of the court was given, and holden not less than twenty-one days after the day on which such decision was given; and
2. The appellant shall, within ten days after the day on which the decision was given, serve notice on the other party and on the clerk of the court of summary jurisdiction, of his intention to appeal, and of the general grounds of such appeal: and
3. The appellant shall, within three days after the day on which he gave notice of appeal, enter into a recognizance before a court of summary jurisdiction, with or without a surety or sureties as the court may direct, conditioned to appear at the said sessions and to try such appeal, and to abide the judgment of the court thereon, and to pay such costs as may be awarded by the court, or the appellant may, if the court of summary jurisdiction thinks it expedient, instead of entering into a recognizance, give such other security, by deposit of money with the clerk of the court of summary jurisdiction or otherwise, as the court deems sufficient; and
4. Where the appellant is in custody a court of summary jurisdiction may, if it seem fit, on the appellant entering into such recognizance or giving such other security as aforesaid, release him from custody; and
5. The court of appeal may adjourn the hearing of the appeal, and upon the hearing thereof may confirm, reverse, or modify the decision of the court of summary jurisdiction, or remit the matter to the court of summary jurisdiction with the opinion of the court of appeal thereon, or make such other order in the matter as the court thinks just.

The court of appeal may also make such order as to costs to be paid by either party as the court thinks just; and

6. Whenever a decision is reversed by the court of appeal, the clerk of the peace shall indorse on the conviction or order appealed against a memorandum that such conviction or order has been quashed, and whenever any copy or certificate of such conviction or order is made, a copy of such memorandum shall be added thereto, and shall be sufficient evidence that the conviction or order has been quashed in every case where such copy or certificate would be sufficient evidence of such conviction or order; and

7. Every notice in writing required by this section to be given by an appellant may be signed by him, or by his agent on his behalf, and may be transmitted in a registered letter by the post in the ordinary way, and shall be deemed to have been served at the time when it would be delivered in the ordinary course of the post.

Provision as to action against person acting in execution of Act.

61. In an action for any act done in pursuance or execution, or intended execution of this Act, or in respect of any alleged neglect or default in the execution of this Act, tender of amends before the action is commenced, may in lieu of or in addition to any other plea be pleaded, if the action was commenced after such tender, or is proceeded with after payment into court of any money in satisfaction of the plaintiff's claim. If the action is commenced after such tender, or is proceeded with after such payment, and the plaintiff does not recover more than the sum tendered or paid respectively, the plaintiff shall not recover any costs incurred after such tender or payment, and the defendant shall be entitled to his costs, to be taxed as between solicitor and client, as from the time of such tender or payment; but this provision shall not affect costs on any injunction in the action.

III.—Miscellaneous.

Continuance of inquisition recorded for ascertaining rents and tolls payable.

62. Every inquisition which, in pursuance of any Act hereby repealed, has been taken for ascertaining the amount of contracts to be performed or rents to be paid in grain or malt, or in any other commodity or thing, or with reference to the measure or weight of any grain, malt, or other commodity or thing, and the amount of any toll rate or duty payable according to any weight or measure in use before the passing of the said Act, and has been enrolled of record in Her Majesty's Court of Exchequer, shall continue in force, and may be given in evidence in any legal proceeding, and the amount ascertained by such inquisition shall, when converted into imperial weights and measures, continue to be the rule of payment in regard to all such contracts, rents, tolls, rates or duties.

Orders in Council.

63. It shall be lawful for Her Majesty in Council from time to time to make Orders for the purposes of this Act, and to revoke and vary any such Order.

All Orders in Council made under this Act shall be published in the London, Edinburgh, and Dublin Gazettes, and shall be forthwith laid before both Houses of Parliament, and shall have full effect as part of this Act.

64. The schedules to this Act, with the notes thereto, shall be construed and have effect as part of this Act. *Effect of schedules.*

65. Where an enactment refers to any Act repealed by this Act, or to any enactment thereof, the same shall be construed to refer to this Act or to the corresponding enactment of this Act. *Construction of Acts referring to repealed enactments.*

Savings and Definitions.

66. Nothing in this Act shall affect the validity of the models of gas holders verified and deposited in the standards department of the Board of Trade in pursuance of the Act of the session of the twenty-second and twenty-third years of the reign of Her present Majesty, chapter sixty-six, intituled "An Act for regulating measures used in sales of gas," and of the Acts amending the same, and the provisions of this Act with respect to Board of Trade standards shall apply to such models; and the provisions of this Act with respect to defining the amount of error to be tolerated in local standards when verified or re-verified, shall apply to defining the amount of error to be tolerated in such copies of the said models of gas holders as are provided by any justices, council commissioners, or other local authority in pursuance of the said Acts. *Saving as to models of gas holders under 22 & 23 Vict. c. 66.*

67. Nothing in this Act shall extend to prohibit, defeat, injure, or lessen the rights granted by charter to the master, wardens, and commonalty of the mystery of Founders of the city of London. *Saving as to rights of the Founders Company.*

68. Nothing in this Act shall prohibit, defeat, injure, or lessen the right of the mayor and commonalty and citizens of the city of London, or of the Lord Mayor of the city of London for the time being, with respect to the stamping or sealing of weights and measures, or with respect to the gauging of wine or oil, or other gaugeable liquors. *Saving as to London.*

69. Nothing in this Act shall extend to supersede, limit, take away, lessen, or prevent the authority which any person or body politic or corporate, or any person appointed at any court leet for any hundred or manor, or any jury or ward inquest, may have or possess for the examining, regulating, seizing, breaking, or destroying any weights, balances, or measures within their respective jurisdictions, and for the purposes of this section the court of burgesses of the city of Westminster shall be deemed to be a body politic, and nothing in this Act shall be deemed to repeal or supersede the Acts relating to that court, or lessen, diminish, or alter the powers of the same. *Act not to abridge the power of the leet jury, &c.*

70. In this Act, unless the context otherwise requires,— The expression "the Summary Jurisdiction Act" means the Act of the session of the eleventh and twelfth years of the reign of Her present Majesty, chapter forty-three, intituled "An Act to facilitate the performance of the duties of justices of the peace out of sessions within England and Wales with respect *Definitions: "Summary Jurisdiction Act:"*

	to summary convictions and orders," inclusive of any Acts amending the same:
"Court of summary jurisdiction:"	The expression "court of summary jurisdiction" means any justice or justices of the peace, metropolitan police magistrate, stipendiary or other magistrate or officer, by whatever name called, to whom jurisdiction is given by the Summary Jurisdiction Act or any Acts therein referred to:
"Quarter Sessions:"	The expression "quarter sessions" includes general sessions:
"Treasury:"	The expression "Treasury" means the Commissioners of Her Majesty's Treasury:
"Person:"	The expression "person" includes a body corporate:
"Stamping:"	The expression "stamping" includes casting, engraving, etching, branding, or otherwise marking, in such manner as to be so far as practicable indelible, and the expression "stamp" and other expressions relating thereto shall be construed accordingly:
"Coin weight:"	The expression "coin weight" means a weight used or intended to be used for weighing coin:
"Weights and Measures Act, 1835."	The expression "Weights and Measures Act, 1835" means the Act of the fifth and sixth years of the reign of King William the Fourth, chapter sixty-three, intituled "An Act to repeal an Act of the fourth and fifth year of His present Majesty relating to weights and measures, and to make other provisions instead thereof."

IV.—APPLICATION OF ACT TO SCOTLAND.

This Act shall apply to Scotland with the following modifications:

Application of imperial weights and measures to tolls, &c.

71. In the application of this Act to Scotland, the expression "rents and tolls" includes all stipends, feu duties, customs, casualties, and other demands whatsoever payable in grain, malt, or meal, or any other commodity or thing.

The fiars prices of all grain in every county shall be struck by the imperial quarter, and all other returns of the prices of grain shall be set forth by the same, without reference to any other measure whatsoever.

Any person who acts in contravention of this provision shall be liable to a fine not exceeding five pounds.

Recovery and application of penalties.

72. All offences under this Act which may be prosecuted, and all fines and forfeitures under this Act which may be recovered on summary conviction, may in Scotland be prosecuted or recovered, with expenses, before the sheriff or sheriff substitute or two or more justices of the peace of the county, or the magistrates of the burgh wherein the offence was committed or the offender resides, at the instance either of the procurator fiscal or of any person who prosecutes.

Every person found liable in Scotland in any fine recoverable summarily under this Act shall, failing payment thereof immediate or within a specified time, as the case may be, and expenses, be liable to be imprisoned for a term not exceeding sixty days, and the

conviction and warrant may be in the form number three of Schedule K of the Summary Procedure Act, 1864. 27 & 28 Vict. c. 53.
All fines and forfeitures so recovered, subject to any payment made to the informer, shall be paid as follows:
  a. To the Queen's and Lord Treasurer's Remembrancer, on behalf of Her Majesty, when the court is the sheriff court:
  b. To the collector of county rates, in aid of the county general assessment, when the court is the justice of the peace court:
  c. To the treasurer of the burgh, in aid of the funds of the burgh, when the court is a burgh court:
  d. To the treasurer of the board of police, or commissioners of police, in aid of the police funds, when the court is a police court.

73. An appeal against a conviction under this Act in Scotland shall be to the Court of Justiciary at the next circuit court, or where there are no circuit courts, to the High Court of Justiciary at Edinburgh, and not otherwise, and such appeal may be made in the manner and under the rules, limitations and conditions contained in the Act of the twentieth year of the reign of King George the Second, chapter forty-three, intituled "An Act for taking away and abolishing heritable jurisdictions in Scotland," or as near thereto as circumstances admit; with this variation, that the appellant shall find caution to pay the fine and expenses awarded against him by the conviction or order appealed from, together with any additional expenses awarded by the court dismissing the appeal. *Appeal.*

74. In the application of this Act to Scotland,— *Definitions as regards Scotland.*
The expression "enter into a recognizance" means grant a bond of caution:
The expression "any court of record" includes the Court of Session and the ordinary sheriff court:
The expression "burgh" shall include royal burgh and parliamentary burgh:
The expression "plaintiff" means pursuer, and the expression "defendant" means defender:
The expression "solicitor" means writer or agent:
The expression "Summary Jurisdiction Act" means the Summary Procedure Act, 1864, inclusive of any Act amending the same. 27 & 28 Vict. c. 53.

75. A sheriff or sheriff substitute shall have the same power in relation to a local comparison of standards, and to the inspection, comparison, seizure and detention of weights and measures, and to entry for that purpose, as is given by this Act to a justice of the peace. *Power of sheriff.*

V.—APPLICATION OF ACT TO IRELAND.

This Act shall apply to Ireland with the following modifications:

76. In Ireland every contract, bargain, sale, or dealing— *Contracts to be made by denominations of imperial*
For any quantity of corn, grain, pulses, potatoes, hay, straw, flax, roots, carcases of beef or mutton, butter, wool, or dead pigs, sold, delivered, or agreed for;

Or for any quantity of any other commodity sold, delivered, or agreed for by weight (not being a commodity which may by law be sold by the troy ounce or by apothecaries weight),

shall be made or had by one of the following denominations of imperial weight; namely,

the ounce avoirdupois;
the imperial pound of sixteen ounces;
the stone of fourteen pounds;
the quarter hundred of twenty-eight pounds;
the half hundred of fifty-six pounds;
the hundredweight of one hundred and twelve pounds; or
the ton of twenty hundredweight;

and not by any local or customary denomination of weight whatsoever, otherwise such contract, bargain, sale, or dealing shall be void:

Provided always, that nothing in the present section shall be deemed to prevent the use in any contract, bargain, sale, or dealing of the denomination of the quarter, half, or other aliquot part of the ounce, pound, or other denomination aforesaid, or shall be deemed to extend to any contract, bargain, sale, or dealing relating to standing or growing crops.

Mode of weighing.

77. In Ireland every article sold by weight shall, if weighed, be weighed in full net standing beam; and for the purposes of every contract, bargain, sale, or dealing the weight so ascertained shall be deemed the true weight of the article, and no deduction or allowance for tret or beamage, or on any other account, or under any other name whatsoever, the weight of any sack, vessel, or other covering in which such article may be contained alone excepted, shall be claimed or made by any purchaser on any pretext whatever under a penalty not exceeding five pounds.

Deductions prohibited.

A proceeding for the recovery of a penalty under this section shall be begun within three months after the offence is committed.

Providing of local standards and sub-standards.

78. 1. The local authority in Ireland shall provide one complete set of local standards for their county or borough; also so many copies in iron or other sufficient material of the local standards.

2. The said copies of the local standards when duly verified as hereinafter mentioned shall be the local sub-standards, and shall be used for the verification of weights and measures brought by the public for verification as if they were local standards.

3. Not less than one set of local sub-standards, and one set of accurate scales shall be provided for each petty sessions district in a county, and not less than two such sets shall be provided for a borough.

4. The local authority shall have the local standards from time to time duly compared and re-verified in manner directed by this Act.

5. The Commissioners of the Dublin Metropolitan Police shall not be under any obligation to provide local standards, but they may, with the assent of the chief secretary or under

secretary to the Lord Lieutenant, procure such sub-standards, scales, and stamps as they think necessary for the purposes of this Act in the district for which they are the local authority.

79. In Ireland, in every year—
   a. in the case of a county, the judge of assize at the first assizes held for the county by inquiry of the foreman of the grand jury; and
   b. in the case of every borough in a county, the recorder of the borough, or, if there be no recorder, the chairman of the quarter sessions for that county, at the quarter sessions held next after the twenty-fifth day of March,

shall inquire whether one complete set of local standards, and a sufficient number of local sub-standards of weights and measures, and a sufficient number of scales and stamps (for verification), have been provided in such county or in such borough.

Inquiry by judge of assize and chairman of quarter sessions as to provision of local standards and sub-standards.

If it appear to the judge or chairman upon such inquiry that the same have not been so provided, he shall forthwith order the proper officer to provide a complete set of local standards, and such sub-standards, scales and stamps as appear to the judge or chairman making the order to be sufficient for the purposes of this Act, and that order shall have the effect in the case of a county of a presentment on the county, and in the case of a borough, of an order on the council of the borough to raise by way of rate, the sum necessary to execute the order, and the said officer shall within three months after he receives the order fully execute the same, and in default shall be liable to a fine not exceeding twenty pounds.

The proper officer shall, in the case of a county, be the treasurer of the county, and in the case of a borough, the town clerk or other proper officer of the borough.

80. Expenses incurred by any member of the Royal Irish Constabulary as an ex-officio inspector of weights and measures in the execution of this Act shall be payable to such inspector by the person acting as treasurer of the local authority of the district on presentation of accounts of such expenses, to be furnished quarterly certified to be correct by the county inspector of the county.

Expenses of ex-officio inspectors.

The secretary of every grand jury being a local authority under this Act shall, at each assizes or presenting term, and the clerk of every other local authority shall once in every year lay before each such grand jury or other local authority an estimate of the sum which may appear to be necessary to meet such expenses until the next assizes or presenting term, or for the ensuing year; and every such grand jury or other local authority shall, without previous application to presentment sessions or other preliminary proceedings, present in advance to the person acting as treasurer the sum specified in such estimate, to be raised and paid out of the local rate; and if the sum so raised proves more than sufficient for the purpose, the balance shall be carried to the credit of the local rate by the person acting as treasurer, and if the sum so raised proves insufficient, the person acting as treasurer shall apply for payment of such expenses any other available funds in his hands.

M

Ex-officio inspectors of weights and measures.

81. Nothing in this Act shall authorise the local authority in Ireland, except the local authority of the borough of Dublin, to appoint inspectors of weights and measures, but such head or other constables in each petty sessions district as may be from time to time selected by the inspector general of constabulary, with the approval of the Lord Lieutenant, shall be ex-officio inspectors of weights and measures under this Act within that district, and shall perform their duties under this Act under the direction of the justices of petty sessions, without fee or reward, and notwithstanding any manorial jurisdiction or claim of jurisdiction within such district :

Provided that if within one month from the date of such selection the justices signify their disapproval of the selection of any head or other constable, another selection shall be made by the same authority, subject to the same conditions, and the inspector general of constabulary shall within three days after any selection has been made in a petty sessions district, give or cause to be given to the clerk of that district notice of such selection, and the clerk shall immediately make known the said selection to the justices of the district.

An ex-officio inspector of weights and measures may exercise, without any authority from a justice of the peace, the powers given by this Act to an inspector of weights and measures having such authority.

In the district in which the commissioners of the Dublin metropolitan police are the local authority under this Act, such of the superintendents, inspectors or acting inspectors of the said police as may be selected by the local authority with the approval of the Lord Lieutenant shall be ex-officio inspectors of weights and measures within the said district.

Custody and use of local standards.

82. The local standards of every county or borough in Ireland shall be in the custody of such sub-inspector of constabulary as may be from time to time appointed for that county or borough by the inspector general of constabulary, with the approval of the Lord Lieutenant.

Such sub-inspector shall, subject to such regulations as the inspector general of constabulary, with the approval of the Lord Lieutenant, from time to time makes, compare with the local standards in his custody, and adjust and verify the local sub-standards sent to him for the purpose, and when the same are correct shall stamp the same with a stamp of verification, and for the purpose of such verification and stamping, and of the verification of local standards, such sub-inspector of constabulary shall be deemed to be an inspector of weights and measures appointed under this Act.

Custody and periodical verification of local substandards.

83. The local sub-standards shall be deposited in the custody of the ex-officio inspector of weights and measures, and shall, at least once in every year, and also at other times when required by the county inspector of constabulary of the county, or by the justices in petty sessions of the county, be compared with the local standards of the county and verified, and when so verified shall, until the expiration of one year or any shorter period at which the next

comparison of the same under this section is made, be deemed to be local sub-standards, and be valid local standards for the purpose of the comparison by way of verification or inspection of weights and measures under this Act.

The sub-standards provided by the commissioners of the Dublin metropolitan police shall be verified by comparison with the local standards of the city of Dublin, as directed by this section, with this qualification, that the said commissioners, and not the county inspector or the justices, shall have authority to require the same to be verified oftener than once a year.

Any person who uses any sub-standard for any purpose other than that authorised by this Act shall be liable to a fine not exceeding five pounds.

84. For the purpose of the prosecution of offences and the recovery of fines under this Act, in Ireland,— *Recovery of fines, &c.*

1. The expression "Summary Jurisdiction Acts" in this Act means, within the police district of Dublin metropolis, the Acts regulating the powers and duties of justices of the peace for such district, or of the police of such district, and elsewhere in Ireland the Petty Sessions (Ireland) Act, 1851, and any Act amending or affecting the same; and *14 & 15 Vict. c. 93.*
2. A court of summary jurisdiction when hearing and determining an information or complaint in any matter arising under this Act shall be constituted within the police district of Dublin metropolis of one of the divisional justices of that district sitting at a police court within the district, and elsewhere of a stipendiary magistrate sitting alone, or with others, or of two or more justices of the peace sitting in petty sessions at a place appointed for holding petty sessions; and
3. Appeals from a court of summary jurisdiction shall lie in the manner and subject to the conditions and regulations prescribed in the twenty-fourth section of the Petty Sessions (Ireland) Act, 1851, and any Acts amending the same. *14 & 15 Vict. c. 93.*

85. In this Act, unless the context otherwise requires, *Definitions.*

The expression "Lord Lieutenant" means the lieutenant or other chief governor or governors of Ireland for the time being:

The expression "treasurer" includes the finance committee and the secretary of the grand jury for the county of Dublin.

VI.—REPEAL.

86. The Acts mentioned in the first part of the Sixth Schedule to this Act are hereby repealed to the extent in the third column of that schedule mentioned; subject to the following qualification, that is to say, that so much of the said Acts as is set forth in the second part of that schedule shall be re-enacted in manner therein appearing, and shall be in force as if enacted in the body of this Act. *Repeal.*

Provided that,—

1. Every inspector appointed in pursuance of any enactment

hereby repealed shall continue in office as if he had been appointed in pursuance of this Act; and

2. Any person holding office as examiner of weights and measures under any enactment repealed by this Act, and not being an inspector of weights and measures within the meaning of this Act, shall continue in office and receive the same remuneration, and have the same powers and duties, and be subject to the same liabilities and to the same power of dismissal as if this Act had not passed.

3. Every notice published in a Gazette in relation to coin weights in pursuance of any enactment hereby repealed, shall continue in force.

4. All weights and measures duly verified and stamped in pursuance of any enactment hereby repealed, shall continue and be as valid as if they had been verified and stamped in pursuance of this Act, and that although such weights or measures could not have been verified and stamped in pursuance of this Act; and all weights and measures which at the commencement of this Act may lawfully be used without being stamped with a stamp of verification or a stamp of their denomination, and which are required by this Act to be stamped with such a stamp, may, notwithstanding they are not so stamped, be used until the expiration of six months after the commencement of this Act, without being subject to be seized or forfeited, and without rendering the person using or having possession of the same subject to any fine.

5. This repeal shall not affect—
   (*a.*) The past operation of any enactment hereby repealed, nor anything duly done or suffered under any enactment hereby repealed; nor
   (*b.*) Any right, privilege, obligation, or liability acquired, accrued, or incurred under any enactment hereby repealed; nor
   (*c.*) Any penalty, forfeiture, or punishment incurred in respect of any offence committed against any enactment hereby repealed; nor
   (*d.*) Any investigation, legal proceeding, or remedy in respect of any such right, privilege, obligation, liability, penalty, forfeiture, or punishment as aforesaid; and any such investigation, legal proceeding, and remedy may be carried on as if this Act had not passed; and

6. This repeal shall not revive any enactment, right, office, privilege, matter, or thing not in force or existing at the commencement of this Act.

SCHEDULES.

FIRST SCHEDULE.

PART I.
IMPERIAL STANDARDS.

Sections 4, 10, 13, 6

The following standards were constructed under the direction of the Commissioners of Her Majesty's Treasury, after the destruction of the former imperial standards in the fire at the Houses of Parliament.

The imperial standard for determining the length of the imperial standard yard is a solid square bar, thirty-eight inches long and one inch square in transverse section, the bar being of bronze or gun-metal; near to each end a cylindrical hole is sunk (the distance between the centres of the two holes being thirty-six inches) to the depth of half an inch, at the bottom of this hole is inserted in a smaller hole a gold plug or pin, about one tenth of an inch in diameter, and upon the surface of this pin there are cut three fine lines at intervals of about the one hundredth part of an inch transverse to the axis of the bar; the measure of length of the imperial standard yard is given by the interval between the middle transversal line at one end and the middle transversal line at the other end, the part of each line which is employed being the point midway between the longitudinal lines; and the said points are in this Act referred to as the centres of the said gold plugs or pins; and such bar is marked "copper 16 oz., tin $2\frac{1}{2}$, zinc 1. Mr. Baily's metal. No. 1 standard yard at $62°\cdot00$ Fahrenheit. Cast in 1845. Troughton & Simms, London."

The imperial standard for determining the weight of the imperial standard pound is of platinum, the form being that of a cylinder nearly $1\cdot35$ inch in height and $1\cdot15$ inch in diameter, with a groove or channel round it, whose middle is about $0\cdot34$ inch below the top of the cylinder, for insertion of the points of the ivory fork by which it is to be lifted; the edges are carefully rounded off, and such standard pound is marked, P.S. 1844, 1 lb.

Sections
5, 35, 64.

PART II.

Parliamentary Copies of Imperial Standards.

The following copies of the standards above-mentioned in part one of this Schedule were constructed at the same time as the above standards. They are of the same construction and form as the above standards, and they are respectively marked and deposited as follows:—

1. One of the copies of the imperial standard for determining the imperial standard yard, being a bronze bar, marked "copper 16 oz., tin 2½, zinc 1. Mr. Baily's metal. No. 2. Standard yard at 61°·94 Fahrenheit. Cast in 1845. Troughton & Simms, London;" and one of the copies of the imperial standard for determining the imperial standard pound marked No. 1., P.C. 1844, 1 lb., have been deposited at the Royal Mint;

2. One other of the copies of the imperial standard for determining the imperial standard yard, being a bronze bar, marked "copper 16 oz., tin 2½, zinc 1. Mr. Baily's metal. No. 3. Standard yard at 62°·10 Fahrenheit. Cast in 1845. Troughton & Simms, London;" and one other of the copies of the imperial standard for determining the imperial standard pound marked No. 2, P.C., 1844, 1 lb., have been delivered to the Royal Society of London;

3. One other of the copies of the imperial standard for determining the imperial standard yard, being a bronze bar, marked "copper 16 oz., tin 2½, zinc 1. Mr. Baily's metal. No. 5. Standard yard at 62°·16 Fahrenheit. Cast in 1845. Troughton & Simms, London;" and one other of the copies of the imperial standards for determining the imperial standard pound marked No. 3, P.C., 1844, 1 lb., have been deposited in the Royal Observatory of Greenwich;

4. The other of the copies of the imperial standard for determining the imperial standard yard, being a bronze bar, marked "copper 16 oz., tin 2½, zinc 1. Mr. Baily's metal. No. 4. Standard yard at 61°·98 Fahrenheit. Cast in 1845. Troughton & Simms, London;" and the other of the copies of the imperial standard for determining the imperial standard pound marked No. 4, P.C., 1844, 1 lb., have been immured in the New Palace at Westminster.

Appendix. 167

SECOND SCHEDULE.

Sections 8, 64.

BOARD OF TRADE STANDARDS.

STANDARDS of the measures and weights following are at the commencement of this Act in use under the direction of the Board of Trade.

MEASURES OF LENGTH.   MEASURES OF CAPACITY.

Denomination of Standard.	Denomination of Standard.
MEASURE OF LENGTH.	MEASURES OF CAPACITY.
100 feet.	Bushel.
66 feet or a chain of 100 links.	Half-bushel.
	Peck.
Rod, pole, or perch.	
10 feet.	Gallon.
6 ,, or 2 yards.	Half-gallon.
5 ,,	Quart.
4 ,,	Pint.
3 ,, or 1 yard.	Half-pint.
2 ,,	Gill.
1 foot.	Half-gill.
1 inch divided into 12 duodecimal, 10 decimal, and 16 binary equal parts.	Quarter-gill.
	MEASURES USED IN THE SALE OF DRUGS.
	Fluid ounces :— 4, 3, 2, 1.
	Fluid drachms :— 4, 3, 2, 1.
	Minims :— 30, 20, 10, 5, 4, 3, 2, 1.

NOTE.—The brass gallon marked "Imperial Standard Gallon, Anno Domini MDCCCXXIV., Anno V G^{IV} Regis," which has a diameter equal to its height, and was made in pursuance of 5 Geo. IV. c. 74, s. 6, and is at the passing of this Act in the custody of the Warden of the Standards, shall be deemed to be a Board of Trade standard for the gallon.

WEIGHTS.

Denomination of Standard.	Denomination of Standard.	Denomination of Standard.
AVOIRDUPOIS WEIGHTS.	TROY BULLION WEIGHTS.	DECIMAL GRAIN WEIGHTS.
56 pounds.	500 ounces.	4,000 grains.
28 ,,	400 ,,	2,000 ,,
14 ,,	300 ,,	1,000 ,,
7 ,,	200 ,,	500 ,,
4 ,,	100 ,,	300 ,,
2 ,,	50 ,,	200 ,,
1 pound.	40 ,,	100 ,,
8 ounces.	30 ,,	50 ,,
4 ,,	20 ,,	30 ,,
2 ,,	10 ,,	20 ,,
1 ounce.	5 ,,	10 ,,
8 drams.	4 ,,	5 ,,
4 ,,	3 ,,	3 ,,
2 ,,	2 ,,	2 ,,
1 dram.	1 ounce.	1 grain.
½ ,,	0·5 ,,	0·5 ,,
240 grains, commonly called 10 pennyweights.	0·4 ,,	0·3 ,,
	0·3 ,,	0·2 ,,
	0·2 ,,	0·1 ,,
120 grains, commonly called 5 pennyweights.	0·1 ,,	0·05 ,,
	0·05 ,,	0·03 ,,
	0·04 ,,	0·02 ,,
72 grains, commonly called 3 pennyweights.	0·03 ,,	0·01 ,,
	0·02 ,,	
	0·01 ,,	
48 grains, commonly called 2 pennyweights.	0·005 ,,	
	0·004 ,,	
	0·003 ,,	
24 grains, commonly called 1 pennyweight.	0·002 ,,	
	0·001 ,,	

Appendix.

COIN WEIGHTS.

Denomination of Coin.	Standard Weight.	
	Imperial Weight.	Metric Weight.
	Grains.	Grams.
Gold:		
Five pound	616·37239	39·94028
Two pound	246·54895	15·97611
Sovereign	123·27447	7·98805
Half-sovereign . . .	61·63723	3·99402
Silver:		
Crown	436·36363	28·27590
Half-crown	218·18181	14·13795
Florin	174·54545	11·31036
Shilling	87·27272	5·65518
Sixpence	43·63636	2·82759
Groat or fourpence . .	29·09090	1·88506
Threepence	21·81818	1·41379
Twopence	14·54545	0·94253
Penny	7·27272	0·47126
Bronze:		
Penny	145·83333	9·44984
Halfpenny	87·50000	5·66990
Farthing	43·75000	2·83495

Sections 18, 64.

THIRD SCHEDULE.

PART I.
METRIC EQUIVALENTS.

TABLE of the Values of the Principal Denominations of Measures and Weights on the Metric System expressed by means of Denominations of Imperial Measures and Weights, and of the values of the Principal Denominations of Measures and Weights of the Imperial system expressed by means of Metric Weights and Measures.

MEASURES OF LENGTH.

Metrical Denominations and Values.		Equivalents in Imperial Denominations.				
—	Metres.	Miles.	Yards.	Feet.	Ins.	Decimals.
Myriametre	10,000	6 or	376	0	11	·9
			10,936	0	11	·9
Kilometre	1,000		1,093	1	10	·79
Hectometre	100		109	1	1	·079
Dekametre	10		10	2	9	·7079
Metre	1		1	0	3	·3708
Decimetre	$\frac{1}{10}$				3	·9371
Centimetre	$\frac{1}{100}$				0	·3937
Millimetre	$\frac{1}{1000}$				0	·0394

MEASURES OF SURFACE.

Metric Denominations and Values.		Equivalents in Imperial Denominations.		
—	Square Metres.	Acres.	Square Yards.	Decimals.
Hectare, i.e. 100 Ares	10,000	2 or	2,280	·3326
			11,960	·3326
Dekare, i.e. 10 Ares	1,000		1,196	·0333
Are	100		119	·6033
Centiare, i.e. $\frac{1}{100}$-Are	1		1	·1960

Appendix.

MEASURES OF CAPACITY.

Metric Denominations and Values.		Equivalents in Imperial Denominations.						
	Cubic Metres.	Quarters.	Bushels.	Pecks.	Gallons.	Quarts.	Pints.	Decimals.
Kilolitre, i.e. 1,000 Litres	1	3	3	2	0	0	0	·77
Hectolitre, i.e. 100 Litres	$\frac{1}{10}$		2	3	0	0	0	·077
Dekalitre, i.e. 10 Litres	$\frac{1}{100}$			1	0	0	1	·6077
Litre	$\frac{1}{1000}$						1	·76077
Decilitre, i.e. $\frac{1}{10}$-Litre	$\frac{1}{10000}$						0	·176077
Centilitre, i.e. $\frac{1}{100}$-Litre	$\frac{1}{100000}$						0	·0176077

WEIGHTS.

Metric Denominations and Values.		Equivalents in Imperial Denominations.					
	Grams.	Cwts.	Stones.	Pounds.	Ounces.	Drams.	Decimals.
Millier	1,000,000	19	5	6	9	15	·04
Quintal	100,000	1	7	10	7	6	·304
Myriagram	10,000		1	8	0	11	·8304
Kilogram	1,000			2 (or 1,5432·3487 grains)		3	4·3830
Hectogram	100				3	8	·4383
Dekagram	10					5	·6438
Gram	1					0	·56438
Decigram	$\frac{1}{10}$					0	·056438
Centigram	$\frac{1}{100}$					0	·0056438
Milligram	$\frac{1}{1000}$					0	·00056438

MEASURES OF LENGTH.

Imperial Measures.	Equivalents in Metric Measures.			
	Millimetre.	Decimetre.	Metre.	Kilometre.
Inch	= 25·39954			
Foot or 12 inches .	..	= 3·04794	= 0·30479	
Yard, or 3 feet, or 36 inches . . .	..	..	= 0·91438	
Fathom, or 2 yards, or 6 feet . . .	..	..	= 1·82877	
Pole or 5½ yards .	..	..	= 5·02911	
Chain, or 4 poles, or 22 yards . . .	..	..	= 20·11644	
Furlong, 40 poles, or 220 yards . . .	..	..	= 201·16437	= 0·20116
Mile, 8 furlongs, or 1,760 yards . .	..	..	= 1,609·31493	= 1·60931

MEASURES OF SURFACE.

Imperial Measures.	Equivalents in Metric Measures.			
	Square Decimetres.	Square Metres.	Ares.	Hectares.
Square inch . .	= 0·06451			
Square foot or 144 square inches .	= 9·28997	= 0·092900		
Square yard, or 9 square feet, or 1,296 sq. inches	= 83·60971	= 0·836097		
Pole or perch, or 30¼ square yards	..	= 25·291939		
Rood, or 40 perches, or 1,210 sq. yards	..	..	= 10·116776	
Acre, or 4 roods, or 4,840 square yds.	..	..	..	= 0·40467
Square mile or 640 acres	..	..	..	= 258·98945

Appendix.

MEASURES OF CAPACITY.

Imperial Measures.	Equivalents in Metric Measures.			
	Decilitres.	Litres.	Dekalitres.	Hectolitres.
Gill	=1·41983	=0·14198		
Pint or 4 gills . .	=5·67932	=0·56793		
Quart or 2 pints .	..	=1·13587		
GALLON or 4 quarts .	..	=4·54346		
Peck or 2 gallons .	..	=9·08692	=0·90869	
Bushel, or 8 gallons, or 4 pecks . .	..	..	=3·63477	
Quarter or 8 bushels	..	..	..	=2·90781

CUBIC MEASURE.

Imperial Measures.	Equivalents in Metric Measures.		
	Cubic Centimetres.	Cubic Decimetres.	Cubic Metres.
Cubic inch	16·38618	..	
Cubic foot or 1,728 cubic inches	..	28·31531	
Cubic yard or 27 cubic feet . .	..	..	0·76451

WEIGHTS.

Imperial Weights.	Equivalents in Metric Weights.			
	Grams.	Dekagrams.	Kilograms.	Millier or Metric Ton.
Grain . . .	= 0·06479895			
Dram . . .	= 1·77185			
Ounce, avoirdupois, or 16 drams, or 437·5 grains . . .	= 28·34954	= 2·83495		
POUND, or 16 ounces, or 256 drams, or 7,000 grains . .	= 453·59265	= 45·35927	= 0·45359	
Hundredweight or 112 lbs. .	..	..	= 50·80238	
Ton or 20 cwt..	..	..	= 1,016·04754	= 1·01605
Ounce, troy, or 480 grains .	= 31·103496	= 3·11035		

174 *Appendix.*

Sections 38, 64.

PART II.

Metric Standards.

List of metric standards in the custody of the Board of Trade at the passing of this Act :—

Measures of Length.

Double metre or 2 metres.
Metre or 1 metre.
Decimetre or 0·1 ,,
Centimetre or 0·01 ,,
Millimetre or 0·001 ,,

Weights.

20, 10, 5, 2 kilograms.
Kilogram.
500, 200, 100, 50, 20, 10, 5, 2, 1 grams.
5, 2, 1 decigrams.
5, 2, 1, 0·5 milligrams.

Measures of Capacity.

20, 10, 5, 2 litres.
Litre.

0·5	litre or	500	cubic centimetres.
0·2	,,	200	,,
0·1	,,	100	,,
0·05	,,	50	,,
0·02	,,	20	,,
0·01	,,	10	,,
0·005	,,	5	,,
0·002	,,	2	,,
0·001	,,	1	,,

Appendix. 175

FOURTH SCHEDULE.

Sections 40, 50, 64.

LOCAL AUTHORITIES.

ENGLAND.

Area.	Local Authority.	Local Rate.
County	The justices in general or quarter sessions assembled.	The county rate.
County of the city of London.	The court of the Lord Mayor and aldermen of the city.	The consolidated rate.
Borough	The mayor, aldermen, and burgesses acting by the council.	The borough fund and borough rate.

SCOTLAND.

Area.	Local Authority.	Local Rate.
County	The justices in general or quarter sessions assembled.	The county general assessment.
Burgh	The magistrates.	The police assessment.

IRELAND.

Area.	Local Authority.	Local Rate.
County	The grand jury acting at any assizes or presenting term.	The presentments to be made by the grand jury.
Such portion of the police district of Dublin metropolis as is without the municipal boundary of the borough of Dublin.	The Commissioners of the Dublin metropolitan police.	The funds applicable to defray the expenses of the Dublin metropolitan police.

Appendix.

IRELAND (continued).

Area.	Local Authority.	Local Rate.
Borough . . .	Town Council.	Rate to be levied by the council, or if the borough is liable to county cess and no rate is levied in the borough, the county cess of the county in which the borough or the larger part thereof is situate.

Notes.

For the purposes of this schedule—

The expression " county," as regards England, does not include a county of a city or a county of a town, but includes every riding, division, or parts of a county having a separate court of quarter sessions. The Soke of Peterborough shall be deemed to be a county, but every other liberty of a county not forming part of the City of London shall be deemed to form part of the county in which the same is situate or which it adjoins, and if it adjoins more than one county, then of the county with which it has the longest common boundary.

The expression " borough," as regards England, means any place for the time being subject to the Municipal Corporation Act, 1835, and any Act amending the same, which has a separate commission of the peace.

The expression " county," as regards Ireland, includes a riding and a county of a city and a county of a town.

The county of Dublin shall be deemed not to include any portion of the police district of Dublin metropolis.

The two constabulary districts of the county of Galway shall respectively be deemed to be counties for the purposes of this Act.

The expression " borough," as regards Ireland, means any borough or town corporate.

In the borough of Dublin the rate to be levied by the council shall mean the improvement rate.

Appendix.

FIFTH SCHEDULE.

Sections 47, 64.

Fees of Inspectors.

The following fees are the maximum fees which, unless altered as authorised by this Act, may be taken by any inspector of weights and measures appointed under this Act.
For comparing and stamping all brass weights:—

	s.	d.
Each half hundredweight	0	9
Each quarter of a hundredweight	0	6
Each stone	0	4
Each weight under a stone to a pound inclusive	0	1
Each weight under a pound	0	0½
Each set of weights of a pound and under	0	2

For comparing and stamping all iron weights, or weights of other descriptions not made of brass:—

	s.	d.
Each half hundredweight	0	3
Each quarter of a hundredweight	0	2
Each stone	0	1
Each weight under a stone	0	0½
Each set of weights of a pound and under	0	2

For comparing and stamping all wooden measures:—

	s.	d.
Each bushel	0	3
Each half bushel	0	2
Each peck, and all under	0	1
Each yard	0	0½

For comparing and stamping all measures of capacity of liquids made of copper or other metal:—

	s.	d.
Each four gallon	0	9
Each two gallon	0	4
Each gallon	0	2
Each half gallon	0	1
Each quart and under	0	0½

Sections 64, 86.

SIXTH SCHEDULE.

First Part.

Enactments repealed.

A description or citation of a portion of an Act is inclusive of the word, section, or other part first or last-mentioned, or otherwise referred to as forming the beginning or as forming the end of the portion described in the description or citation.

Portions of Acts which have already been specifically repealed are in some instances included in the repeal in this schedule, in order to preclude henceforth the necessity of looking back to previous Acts.

Session and chapter.	Title or short title of Act.	Extent of repeal.
31 Edw. 3. st. 1.	The statute made at Westminster on the Monday next after the feast of Easter, in the thirty-first year, statute the first.	Chapter two.
6 Anne, c. 11. (5 & 6 Anne, c. 8. in Ruffhead.)	An Act for the union of the two kingdoms of England and Scotland.	Article seventeen.
15 Geo. 2. c. 20.	An Act to prevent the counterfeiting of gold and silver lace, and for settling and adjusting the proportions of fine silver and silk, and for the better making of gold and silver thread.	Section five.
35 Geo. 3. c. 102.	An Act for the more effectual prevention of the use of defective weights, and of false and unequal balances.	The whole Act.

Appendix.

Session and chapter.	Title or short title of Act.	Extent of repeal.
36 Geo. 3. c. 85.	An Act for the better regulation of mills.	Section one from "and any per-" "son or persons" "appointed" down to "with" "respect to" "weights and" "balances," and from "and" "every miller" "or other per-" "son as afore-" "said, in whose" "mill shall be" "found any" "weight or" "weights" to the end of the section.
37 Geo. 3. c. 143.	An Act to explain and amend an Act made in the thirty-fifth year of the reign of His present Majesty, intituled "An Act for the more effec-" "tual prevention of the use" "of defective weights and of" "false and unequal balances."	The whole Act.
55 Geo. 3. c. 43.	An Act for the more effectual prevention of the use of false and deficient measures.	The whole Act.
5 Geo. 4. c. 74.	An Act for ascertaining and establishing uniformity of weights and measures.	The whole Act, except section twenty-five.
6 Geo. 4. c. 12.	An Act to prolong the time of the commencement of an Act of the last session of Parliament for ascertaining and establishing uniformity of weights and measures, and to amend the said Act.	The whole Act.

Session and chapter.	Title or short title of Act.	Extent of repeal.
5 & 6 Will. 4. c. 63.	An Act to repeal an Act of the fourth and fifth year of His present Majesty relating to weights and measures, and to make other provisions instead thereof.	The whole Act.
16 & 17 Vict. c. 29.	An Act for regulating the weights used in sales of bullion.	The whole Act.
16 & 17 Vict. c. 79.	An Act for making sundry provisions with respect to municipal corporations in England.	Section five.
18 & 19 Vict. c. 72.	An Act for legalizing and preserving the restored standards of weights and measures.	The whole Act.
22 & 23 Vict. c. 56.	An Act to amend the Act of the fifth and sixth years of King William the Fourth, chapter sixty-three, relating to weights and measures.	The whole Act.
23 & 24 Vict. c. 119.	An Act to amend the law relating to weights and measures in Ireland.	The whole Act.
24 & 25 Vict. c. 75.	An Act for amending the Muncipal Corporations Act.	Section six.
25 & 26 Vict. c. 76.	The Weights and Measures (Ireland) Amendment Act, 1862.	The whole Act, except section two, and Part three and so much of Part four as relates to Part three.
25 & 26 Vict. c. 102.	The Metropolis Management Amendment Act, 1862.	Section one hundred and one.
27 & 28 Vict. c. 117.	The Metric Weights and Measures Act, 1864.	The whole Act.
29 & 30 Vict. c. 82.	An Act to amend the Acts relating to the standard weights and measures, and to the standard trial pieces of the coin of the realm.	The whole Act.

Session and chapter.	Title or short title of Act.	Extent of repeal.
30 & 31 Vict. c. 94.	An Act to provide for the inspection of weights and measures and to regulate the law relating thereto, in certain parts of the police district of Dublin metropolis.	The whole Act.
33 & 34 Vict. c. 10.	The Coinage Act, 1870.	Section seventeen, from the beginning of the section down to "weight of and" "for weighing" "such coin," and from "all" "weights which" "are not less in" "weight" to the end of the section.

Second Part.

Enactments re-enacted.

5 & 6 Will. 4, c. 63, s. 9.

All coals, slack, culm, and cannel of every description shall be sold by weight, and not by measure. Every person who sells any coals, slack, culm, or cannel of any description by measure, and not by weight, shall be liable on summary conviction to a fine not exceeding forty shillings for every such sale. *Sale of coals by weight and not by measure.*

5 & 6 Will. 4, c. 63, s. 26.

In Ireland, in every city or town, not being a county of itself, every person, persons, or body corporate exercising the privilege of appointing a weigh-master, shall supply him with accurate scales, and with an accurate set of copies of the *Supply of weigh-masters in Ireland with scales, and* local standards, and in default shall be liable on summary conviction to a fine of twenty pounds, and the accuracy of such set of copies shall be certified under the hand of some inspector of weights and measures. They shall also, once at least in every five years, cause such copies to be readjusted by comparison with some local standards which have been verified by the Board of Trade, and in default shall be liable on summary conviction to a fine of five pounds.

<small>copies of local standards.</small>

Such set of copies shall for the purpose of comparison and verification be considered local standards, and shall be used for no other purpose whatever, and if they are so used, the person using the same shall be liable on summary conviction to a fine of five pounds.

22 & 23 Vict. c. 56, ss. 6, 8, 12.

The owners or managers of any public market in Great Britain where goods are exposed or kept for sale, shall provide proper scales and balances, and weights and measures or other machines, for the purpose of weighing or measuring all goods sold, offered, or exposed for sale in any such market, and shall deposit the same at the office of the clerk or toll collector of such market, or some other convenient place, and shall have the accuracy of all such scales and balances and weights and measures or other machines tested at least twice in every year by the inspector of weights and measures of and for the county, borough, or place where the market is situate;

<small>Owners of markets to provide scales, &c.</small>

All expense attending the purchase, adjusting, and testing thereof shall be paid out of the moneys collected for tolls in the market;

Such clerk or toll collector shall at all reasonable times, whenever called upon so to do, weigh or measure all goods which have been sold, offered, or exposed for sale in any such market, upon payment of such reasonable sum as may from time to time be decided upon by the said owners or managers, subject to the approval and revision of the justices in general or quarter sessions assembled if such market be in England, or of the sheriff if it be in Scotland;

For every contravention of this section the offender shall be liable, on summary conviction, to a fine not exceeding five pounds.

22 & 23 Vict. c. 56, ss. 7, 8, 12.

Every clerk or toll collector of any public market in Great Britain, at all reasonable times, may weigh or measure all goods sold, offered, or exposed for sale in any such market; and if upon such weighing or measuring any such goods are found

<small>Power to clerks of markets to inspect goods sold, &c., and</small>

deficient in weight or measure or otherwise contrary to the provisions of this Act, such clerk or toll collector shall take the necessary proceedings for recovering any fine, to which the person selling, offering, or exposing for sale, or causing to be sold, offered, or exposed for sale, such goods, is liable, and the court convicting the offender may award out of the fine to such clerk or toll collector such reasonable remuneration as to the court seems fit. *if weighing found deficient to summon the offender.*

For every offence against or disobedience to this section the offender shall be liable on summary conviction to a fine not exceeding five pounds.

ANNO VICESIMO QUARTO & VICESIMO QUINTO
VICTORIÆ REGINÆ.

CAP. LXXVIII.

An Act to repeal certain Enactments relating to nominating and appointing the Householders of *Westminster* to serve as Annoyance Jurors, and to make other Provisions in lieu thereof.

[6th *August* 1861.]

27 Eliz. c. 17, 29 G. 2, c. 25, 31 G. 2, c. 17.

WHEREAS an Act was passed in the Twenty-seventh Year of the Reign of Queen *Elizabeth*, intituled *An Act for the good Government of the City or Borough of* Westminster *in* Middlesex, which Act, as amended by the Twenty-ninth of King *George* the Second, Chapter Twenty-five, and Thirty-first of King *George* the Second, Chapter Seventeen, appointed, among other things, an Annoyance Jury, to inspect Annoyances, Obstructions, and Weights and Measures of Traders in the said City and Borough: And whereas it is expedient that so much of the said Act and such amended Acts as relate to the Appointment and Duties of such Annoyance Jurors should be repealed, and other Provisions made instead thereof: Be it therefore enacted by the Queen's most Excellent Majesty, by and with the Advice and Consent of the Lords Spiritual and Temporal, and Commons, in this present Parliament assembled, and by the Authority of the same, as follows;

Repeal of part of recited Acts after 29 Sept. 1861.

1. From and after the Twenty-ninth Day of *September* One thousand eight hundred and sixty-one, so much of the said Act and amended Acts as relates to the Appointment of Annoyance Jurors shall be and the same is hereby repealed.

Not to extend to Offences committed before the passing of this Act.

2. Provided always, That nothing herein contained shall extend or be construed to extend to interfere with any Acts done or Appointments made under the Authority of the said recited Acts, or to prevent the suing for or Recovery of any Penalty incurred by any Offence committed against the Provisions of the said recited Acts previous to the repeal thereof in and by this Act.

Dean and Court of Burgesses to appoint, remunerate, suspend, or

3. On and after the Twenty-ninth Day of *September* One thousand eight hundred and sixty-one, the Appointment of such Annoyance Jurors shall cease, and the Dean of the Collegiate Church of *Saint Peter Westminster* for the Time being, or the High Steward of the City and Liberty of *Westminster* for the Time being, or his lawful Deputy, shall and may, from Time to Time, as Circumstances may

Appendix.

require, call a Meeting or Meetings of the Court of Burgesses of the City and Liberty of *Westminster*, at which Court the said Dean, or the High Steward or his Deputy, or one of the Chief Burgesses and Four of the Burgesses, shall be present; and such Court so constituted shall exercise the powers by this Act given to the Court of Burgesses, and shall and may and is hereby required to appoint One or more Inspectors of Weights and Measures, who shall hold the Office during the Pleasure of the said Court, which is hereby empowered to suspend or dismiss any Inspector so appointed, and to appoint other Inspectors as Occasion may require, and shall direct what reasonable Remuneration shall be paid to every Inspector for the Discharge of such Duties as he is ordered by the said Court of Burgesses to perform, within the limits of its Jurisdiction, for preventing persons dealing by unlawful Weights, Balances, or Measures within the said City or Liberty of *Westminster*. — *discharge of Inspectors.*

4. Provided always, That every Inspector under this Act, before he enters upon the Execution of his Office, shall take an oath to the Effect following, which Oath the said Court of Burgesses is hereby empowered to administer: — *Oath to be taken by Inspector.*

" I A.B. do swear that I will faithfully, impartially, and honestly, according to the best of my Skill, Judgment, and Ability, execute the Powers and Duties of an Inspector of Weights and Measures under an Act passed in the   Year of the Reign of Queen Victoria, intituled 'An Act to repeal certain Enactments relating to nominating and appointing the Householders of Westminster to serve as Annoyance Jurors, and to make other Provisions in lieu thereof,' and that I will execute those Powers and Duties without Hatred or Malice, Fear, Favour, or Affection.

" So help me GOD."

5. Every Inspector under this Act shall and may, with or without One or more Person or Persons acting by his or their Authority, at all seasonable Times during the Hours of Business in the Day or Night, enter any House, Shop, Warehouse, Building, or Yard within the said City and Liberty in the Occupation of or used by any Person who deals by Weight or Measure, and search for, take, and examine all Weights, Measures, Balances, Steelyards, and Weighing Machines there found and being, and if any of the same be unlawful, fraudulent, or defective, he may and is hereby directed and required to seize, keep, and detain the same, and to cause to be summoned the person so offending before the said Court of Burgesses, which, in default of the Appearance of such Person, or after hearing such Person or any other One Individual who may appear on his Behalf, shall, on Proof thereof on Oath, fine such Person so offending in any Sum not exceeding Five Pounds for any One Offence, and the unlawful or defective Weights, Measures, Balances, Steelyards, and Weighing Machines shall thereupon be forfeited to the said Court and destroyed. — *Inspectors to visit shops and warehouses; power to seize weights and summon offenders; maximum of fines fixed.*

6. Every Inspector under this Act may, with or without One or more Persons acting by his Authority, at all seasonable times in the Day or Night, search for, take, and examine all Weights, — *Inspector to inspect weights and measures*

of persons in the streets; power to break unjust weights and summon offenders; maximum of fines fixed.	Measures, Balances, Steelyards, and Weighing Machines in the Possession of any Person selling, offering, or exposing for Sale any Goods on any open Ground, or in any public Street, Lane, Thoroughfare, or place within the said City and Liberty of *Westminster;* and if upon such Examination, any such Weights, Measures, Balances, Steelyards, and Weighing Machines be found unlawful, fraudulent, or defective, or shall be used in a fraudulent Manner, the same shall thereupon forthwith be forfeited to the said Court, and be seized, detained, and destroyed; and any Person using or having in his Possession any such unlawful, fraudulent, or defective Weights, Measures, Balances, Steelyards, and Weighing Machines, or using any Weights, Measures, Balances, Steelyards, or Weighing Machines in a fraudulent manner, shall be summoned before the said Court of Burgesses, which, in default of the Appearance of such Person, or after hearing such person, or any other One Individual who may appear on his Behalf, shall, on Proof thereof on oath, fine such Person in any sum not exceeding Five Pounds for any One Offence.
Power to Court to summon and examine witnesses.	7. The Court of Burgesses may summon Witnesses to give Evidence before that Court touching any Matters arising under the Fifth and Sixth Sections of this Act, or either of them, and may examine those Witnesses on Oath, and may do all things necessary for the due and proper Hearing and Determination of any of the said matters so arising as aforesaid.
Summons to be under the Seal of this Court. Service of summonses. Penalty on witnesses for not attending and giving Evidence.	8. Every Summons under this Act shall be issued by the Town Clerk under the Common Seal of the said Court of Burgesses. 9. Every summons under this Act may be served upon the Person to whom it is directed by delivering the same to such Person personally, or by leaving the same with some Person for him at his last or most usual Place of Business or Abode. 10. Any Person summoned as a Witness to give Evidence before the said Court of Burgesses touching any Matters arising under the said Fifth and Sixth Sections of this Act or either of them, who shall neglect or refuse to appear at the Time and Place for that Purpose appointed, and who shall not make such reasonable Excuse for such Neglect or Refusal as shall be admitted and allowed by the said Court of Burgesses, or who appearing shall refuse to be examined on Oath or Affirmation and give Evidence, shall, on Conviction by the said Court of Burgesses, forfeit and pay to the said Court of Burgesses a Fine not exceeding Five Pounds for every such Offence.
Forms of summons and conviction.	11. Any Summons or Conviction under this Act may be in the Form given in the Schedule to the Act passed in the Twelfth Year of the Reign of Her Majesty, Chapter Forty-three, so far as any Form of Summons or Conviction therein may be applicable to the particular case, and with such alterations or additions as the Circumstances of each Case may require, and every such Form or any Form to the like Effect shall be deemed good, valid, and sufficient in the Law.
Penalties for obstructing Inspector.	12. Every Person who shall abuse or insult any such Inspector when in the execution of his office, or shall in any way obstruct

the execution of the said office, shall be liable to a penalty not exceeding forty shillings.

13. Every inspector, and every Person acting under his authority, who shall ask, demand, or take any Sum of Money or other Gratuity or Reward whatsoever for or under Pretence of excusing any Person or Persons, or for not summoning any Person or Persons for any Offences committed under this Act, or shall otherwise misconduct himself in the Execution of his Office, shall be liable to a Penalty not exceeding Five Pounds. *Penalties for misconduct of Inspector.*

14. All Fines imposed by the said Court of Burgesses under this Act shall and may be levied and recovered in the like Manner as the Fines and Amerciaments set or imposed by the said recited Acts are thereby directed to be levied and recovered; and all Sums so recovered shall be applied and disposed of in the Manner following; that is to say, the High Bailiff of *Westminster* or his deputy for the Time being shall be and he is hereby entitled to one Moiety thereof, and shall receive and take the same to his own Use, and the other Moiety thereof shall be taken and applied by the said Court of Burgesses to pay the necessary Charges and Expenses that shall attend the Execution of this Act. *Fines to be paid to High Bailiff and Court of Burgesses.*

15. The Penalties imposed by the Twelfth and Thirteenth Sections of this Act shall be applied and recovered in the same way as if the Offences created and Penalties imposed by those Sections had, at the passing of an Act passed in the Third Year of the Reign of Her Majesty, Chapter Forty-seven, been created and imposed by and those Sections had been enacted in the Fifty-fourth Section of that Act. *Penalties for obstructing and for misconduct of Inspector recoverable under the Metropolitan Police Act.*

16. If any Vacancy shall at the passing of this Act exist or hereafter occur in the Office of sizing and sealing Weights and Measures under the said Acts or either of them, every Person appointed to fill that office shall hold the same during the Pleasure of the said Court of Burgesses, which may suspend or dismiss every such Person, and appoint others, as Occasion requires. *Office of Sizer and Sealer to be held during the pleasure of the Court of Burgesses.*

17. Nothing in this Act contained shall, except so far as by this Act expressly provided, extend or be construed to extend to interfere with the appointment of any officer by the Court of Burgesses, or with sealing, sizing, stamping or marking of any Weights or Measures, or with the Fees for sealing, sizing, stamping or marking such Weights and Measures payable before the passing of this Act; and this Act shall be construed and taken together with the said recited Acts, and the said Acts and this Act shall, so far as the Provisions of the same are respectively consistent, be read together as One Act. *Reserving rights of the Court of Burgesses.*

FORMS OF WARRANT FOR INSPECTORS.

_____ to wit. WHEREAS an Act was passed in the forty-second year of the reign of Her Majesty Queen Victoria, intituled " The Weights and Measures Act, 1878," and WHEREAS by the forty-eighth section of the said Act, it is enacted that every inspector of weights and measures under that Act, authorised in writing under the hand of a justice of the peace, also every justice of the peace, may, at all reasonable times, inspect all weights, measures, scales, balances, steelyards and weighing machines, within his jurisdiction, which are used or in the possession of any person, or on any premises for use for trade, and may compare every such weight and measure with some local standard, and may seize and detain any weight, measure, scale, balance, or steelyard, which is liable to be forfeited in pursuance of that Act, and may, for the purpose of such inspection, enter any place, whether a building or in the open air, whether open or enclosed, where he has reasonable cause to believe that there is any weight, measure, scale, balance, steelyard, or weighing machine which he is authorised by that Act to inspect;

Now I,   , Esquire, one of Her Majesty's justices of the peace for the   of   , do hereby authorise   inspector of weights and measures, duly appointed in accordance with the provisions of the Weights and Measures Act, 1878, for the district   , in the   of   , at all reasonable times to inspect all weights, measures, scales, balances, steelyards, and weighing machines within his jurisdiction, which are used or in the possession of any person, or on any premises for use for trade, and to compare every such weight and measure with some local standard, and to seize and detain any weight, measure, scale, balance, or steelyard, which is liable to be forfeited in pursuance of the said Weights and Measures Act, 1878, and, for the purpose of such inspection to enter any place, whether a building or in the open air, whether open or enclosed, where he has reasonable cause to believe that there is any weight, measure, scale, steelyard, or weighing machine which he is authorised by the Weights and Measures Act, 1878, aforesaid, to inspect.

Given under my hand this   day of   , one thousand eight hundred and   .

OR,

_____ to wit. To   , inspector of
weights and measures for the   of   , in the
          of   , and to all others whom it may
concern.

WHEREAS you, the said   , have been duly
appointed and now are an inspector of weights and measures for
the   of   , in the said county of   ,
in accordance with the provisions of the Weights and Measures
Act, 1878;

Now I,   , Esquire, one of Her Majesty's
justices of the peace in and for the said county, in pursuance of
the forty-eighth section of the Weights and Measures Act, 1878,
aforesaid, do hereby authorise you the said   as
such inspector as aforesaid, at all reasonable times to inspect all
weights, measures, scales, balances, steelyards, and weighing
machines within your jurisdiction which are used or in the
possession of any person or on any premises for use for trade, and
to compare every such weight and measure with some local
standard, and to seize and detain any weight, measure, scale,
balance, or steelyard which is liable to be forfeited in pursuance
of the Weights and Measures Act, 1878, aforesaid, and for the
purpose of such inspection to enter any place, whether a building
or in the open air, whether open or enclosed, where you have
reasonable cause to believe that there is any weight, measure,
scale, balance, steelyard, or weighing machine which you are
authorised by the Weights and Measures Act, 1878, aforesaid to
inspect, and for so doing this shall be your sufficient warrant and
authority.

Given under my hand this   day of   ,
18   .

FORM OF CERTIFICATE OF STAMPING.

No.

County of _____
District of _____

_____ 18   .

CERTIFICATE OF WEIGHTS AND MEASURES STAMPED FOR
Mr. _____

	£.	s.	d.
Brass Weights.			
Half cwts.			
Quarter do.			
Stone			
Under a stone to a pound inclusive			
Under a pound			
Set of a pound and under . .			
Iron Weights.			
Half cwts.			
Quarter do.			
Stone			
Under a stone			
Set of a pound and under . .			
Dry Measures.			
Bushel			
Half bushel			
Peck and under			
Yard Measure			
Liquid Measures.			
Four gallon			
Two gallon			
Gallon			
Half gallon			
Quart and under			
£			

_____,
Inspector of Weights and Measures.

LIST OF OFFENCES UNDER THE WEIGHTS AND MEASURES ACT, 1878.

41 & 42 VICT. c. 49.

1. Selling by any denomination of weight or measure other than one of the imperial weights or measures, or some multiple or part thereof. *(Selling by other than imperial weights and measures.)*

    Fine not exceeding forty shillings for every such sale.

2. All articles sold by weight shall be sold by avoirdupois weight, except that— *(Sale by avoirdupois weight, with exceptions.)*
    (a) Gold and silver, and articles made thereof, including gold and silver thread, lace, or fringe, also platinum, diamonds, and other precious metals or stones, may be sold by the ounce troy, or by any decimal parts of such ounce; and
    (b) Drugs, when sold by retail, may be sold by apothecaries' weight.

    Penalty for acting in contravention of this section: Fine not exceeding five pounds.

3. Printing, making, or publishing any return, price list, or price current, in which the denomination of weights and measures quoted or referred to denotes or implies a greater or less weight or measure than is denoted or implied by the same denomination of the imperial weights or measures. *(Publishing price lists, &c., showing other denominations of weights and measures than imperial.)*

    Fine not exceeding ten shillings for every copy of every such return, price list, or price current.

4. USING OR HAVING IN POSSESSION FOR USE FOR TRADE a weight or measure, which is not of the denomination of some Board of Trade standard. *(Unauthorised weights and measures.)*

    Fine not exceeding five pounds, or, if a second offence, ten pounds. Weight or measure may be seized and forfeited.

5. USING OR HAVING IN POSSESSION FOR USE FOR TRADE any weight, measure, scale, balance, steelyard, or weighing machine, which is false or unjust. *(Unjust weights, measures, balances, &c.)*

    Fine not exceeding five pounds, or, if a second offence, ten pounds. Contracts, &c., made by same void. Weight, measure, scale, balance, or steelyard may be seized and forfeited.

Fraud in use of weights, measures, balances, &c.	6. Wilfully committing any fraud in the using of any weight, measure, scale, balance, steelyard, or weighing machine.

> *Fine not exceeding five pounds, or, if a second offence, ten pounds. Persons party to the fraud also liable. Weight, measure, scale, balance, or steelyard may be seized and forfeited.*

Making and selling unjust weights and measures.	7. Wilfully or knowingly making or selling, or causing to be made or sold, any false or unjust weight, measure, scale, balance, steelyard, or weighing machine.

> *Fine not exceeding ten pounds, or, if a second offence, fifty pounds.*

Using unstamped measures and weights.	8. USING OR HAVING IN POSSESSION FOR USE FOR TRADE any measure or weight not stamped by an inspector.

> *Fine not exceeding five pounds, or, if a second offence, ten pounds. Contracts, &c., made by same void. Measure or weight may be seized and forfeited.*

Lead and pewter weights to be cased.	9. A weight made of lead or pewter shall not be stamped or used for trade, unless it be wholly and substantially cased with brass, copper, or iron, and legibly stamped or marked "cased." A plug of lead or pewter, the insertion of which into a weight is *bonâ fide* necessary for the purpose of adjusting it, and affixing the stamp of verification, is allowed.

> *Penalty for disobeying the provisions of this section: Fine not exceeding five pounds, or, if a second offence, ten pounds.*

Coin weights to be stamped by the Board of Trade.	10. Every coin weight, not less in weight than the weight of the lightest coin for the time being current, shall be verified and stamped by the Board of Trade.

> *Penalty for using an unstamped coin weight, for determining the weight of gold and silver coin of the realm: Fine not exceeding fifty pounds.*

Forging stamp on weights and measures; and uttering same, &c.	11. Forging or counterfeiting any stamp used for the stamping of any weight or measure.

> *Fine not exceeding fifty pounds.*

Wilfully increasing or diminishing a stamped weight.

> *Fine not exceeding fifty pounds.*

Knowingly using, selling, uttering, disposing of, or exposing for sale, any measure or weight with a forged or counterfeit stamp thereon, or any stamped weight which has been wilfully increased or diminished.

Fine not exceeding ten pounds. All weights and measures with any such forged stamp shall be seized and forfeited.

12. Neglecting or refusing to produce for inspection all weights, measures, scales, balances, steelyards, and weighing machines on the premises; or refusing to permit a justice or inspector to examine the same; or obstructing the entry of a justice or inspector; or otherwise obstructing or hindering a justice or inspector. *Obstructing inspectors, &c.*

Fine not exceeding five pounds, or, if a second offence, ten pounds.

13. Any inspector illegally stamping a weight or measure, or guilty of any breach of duty or misconduct. *Misconduct of inspectors.*

Fine not exceeding five pounds for each offence.

For offences peculiar to SCOTLAND and IRELAND, see Chapter XI.

ORDERS IN COUNCIL.

AT the Court at *Osborne House, Isle of Wight,* the 4th day of February, 1879.

PRESENT,

The QUEEN'S Most Excellent Majesty in Council.

WHEREAS by "The Weights and Measures Act, 1878," it is (among other things) provided that the Board of Trade shall from time to time cause such new denominations of Standards, being either equivalent to or multiples or aliquot parts of the imperial weights and measures ascertained by the said Act, as appear to them to be required in addition to those mentioned in the second schedule to the said Act, to be made and duly verified, and that those new denominations of Standards, when approved by Her Majesty in Council, shall be Board of Trade Standards in like manner as if they were mentioned in the said schedule:

And whereas it has been made to appear to the Board of Trade that a new denomination of Standard weight of 100 pounds, being a multiple of the imperial weight of one pound ascertained by the said Act, is required, and they have caused

O

the same to be made and duly verified and deposited in their custody:

And whereas the Board of Trade have given to the said new Standard weight the denomination of "Cental or new Hundredweight":

Now, therefore, Her Majesty, by virtue of the power vested in Her by the said Act, by and with the advice of Her Privy Council, is pleased to approve of the "Cental or new Hundredweight" as a new denomination of Standard, and doth direct that the same shall be a Board of Trade Standard in like manner as if it was mentioned in the second schedule to "The Weights and Measures Act, 1878."

C. L. Peel.

AT the Court at *Osborne House, Isle of Wight*, the 4th day of February, 1879.

PRESENT,

The QUEEN'S Most Excellent Majesty in Council.

WHEREAS by "The Weights and Measures Act, 1878," it is (among other things) provided that it shall be lawful for Her Majesty from time to time, by Order in Council, to define the amount of error to be tolerated in local standards when verified or re-verified by the Board of Trade:

And whereas it is by the same Act also provided, that the provisions of the said Act with respect to defining the amount of error to be tolerated in local standards when verified or re-verified, shall apply to defining the amount of error to be tolerated in any copies of the models of gasholders verified and deposited in the Standards Department of the Board of Trade as are provided by any justices, council, commissioners, or other local authority in pursuance of the Act of the session of the twenty-second and twenty-third years of the reign of Her present Majesty, chapter sixty-six, intituled "An Act for regulating measures used in the sales of gas," and of the Acts amending the same:

And whereas it has been made to appear to Her Majesty that the scale of errors contained in the schedule hereto should hereafter be tolerated in local standards when verified or re-verified by the Board of Trade:

Now, therefore, Her Majesty, by virtue of the power vested in Her by the said Act, by and with the advice of Her Privy Council, is pleased to define that the amount of errors set forth in the schedule hereto may from and after the date of this Order be tolerated in local standards when verified or re-verified by the Board of Trade.

C. L. Peel.

Appendix. 195

SCHEDULE.

SCALE OF ERRORS to be tolerated in Local Standards when verified or re-verified by the Board of Trade.

1. LOCAL STANDARD WEIGHTS.

Denomination.	Amount of error tolerated in excess. No error in deficiency allowed.
Avoirdupois Weights: 56 lb. 28 ,, 14 ,,	} 5 grains.
7 ,, 4 ,, 2 ,,	} 2 grains.
1 lb. 8 oz. 4 ,,	} 0·5 grain.
2 ,, 1 ,, 8 drams	} 0·2 grain.
4 ,, to ½ dram	} 0·1 grain.
240 grains, commonly called 10 pennyweights	} 0·1 grain.
120 grains, commonly called 5 pennyweights	} 0·05 grain.
72 grains, commonly called 3 pennyweights	} 0·04 grain.
48 grains, commonly called 2 pennyweights	} 0·03 grain.
24 grains, commonly called 1 pennyweight	} 0·02 grain.

1. LOCAL STANDARD WEIGHTS—*continued*.

Denomination.	Amount of error tolerated in excess. No error in deficiency allowed.
Troy Bullion Weights:	
500 oz.	
400 ,,	
300 ,,	2 grains.
200 ,,	
100 ,,	
50 ,,	
40 ,,	0·5 grain.
30 ,,	
20 ,,	
10 oz.	
5 ,,	
4 ,,	0·1 grain.
3 ,,	
2 ,,	
1 oz.	
0·5 ,,	
0·4 ,,	0·5 grain.
0·3 ,,	
0·2 ,,	
0·1 ,,	
0·05 ,,	
0·04 ,,	
0·03 ,,	0·02 grain.
0·02 ,,	
0·01 ,,	
0·005 ,,	
0·004 ,,	
0·003 ,,	0·01 grain.
0·002 ,,	
0·001 ,,	
Decimal Grain Weights:	
4000	
2000	0·1 grain.
1000	

Appendix.

1. LOCAL STANDARD WEIGHTS—*continued.*

Denomination.	Amount of error tolerated in excess. No error in deficiency allowed.
Decimal Grain Weights: 500, 300, 200, 100	0·04 grain.
50, 30, 20, 10	0·03 grain.
5, 3, 2, 1	0·02 grain.
0·5, 0·3, 0·2, 0·1	0·01 grain.
0·05, 0·03, 0·02, 0·01	No appreciable error allowed.

2. LOCAL STANDARD MEASURES OF CAPACITY.

Denomination.	Amount of error tolerated in excess or in deficiency. Grains weight of water as measured by a graduated glass tube.
Measures of Capacity:	
Bushel	256 grains.
Half-bushel	128 ,,
Peck	64 ,,
Gallon	32 ,,
Half-gallon	16 ,,

2. Local Standard Measures of Capacity—*continued*.

Denomination.	Amount of error tolerated in excess or in deficiency. Grains weight of water as measured by a graduated glass tube.
Measures of Capacity: Quart, Pint, Half-pint	8 grains.
Gill, Half-gill, Quarter-gill	4 ,,
Measures used in the sale of drugs:	
4, 3, 2, 1 fluid ounces	4 ,,
4, 3, 2, 1 fluid drachms	2 ,,
30, 20, 10, 5, 4, 3, 2, 1 minims	No appreciable error allowed.

3. Local Standard Measures of Length.

Denomination.	Amount of error tolerated in excess or in deficiency.
100 feet, 66 feet or a chain of 100 links, Rod, pole, or perch	$\frac{1}{10}$ or 0·1 inch.
10 feet, 6 ,, 5 ,, 4 ,,	$\frac{2}{100}$ or 0·02 inch.
3 ,, or 1 yard, 2 ,, 1 foot	$\frac{1}{100}$ or 0·01 inch.
1 inch and sub-divisions of the inch	No appreciable error allowed.

Appendix.

4. Local Standard Gas Measurers.

Denomination.	Amount of error tolerated.
10 cubic feet gas-holder . . 5 ,, ,, ,, . . 1 ,, ,, ,, . .	$\frac{1}{4}$ or 0·25 per centum in excess and in deficiency.
Test gas-meters . . .	$\frac{1}{4}$ or 0·25 per centum fast or slow.

INDEX.

Acre of land, definition of, 2.
Actions against inspectors, 78.
Adjournment of hearing informations, 67.
Adjusting weights and measures, 52.
Administration, 15.
Apothecaries' weight, 4, 38.
Appeal, proceedings upon, 73–78.
Assize of bread and ale, 29.
Attendance of defendant, how enforced, 66.
——————— witness, ,, 68.
Aulnage of cloths, 30.
Auncel weight, 30.
Averia, 38.
Avoirdupois weight, 3.
——————————, all sales to be by, 32.
——————————, origin of, 37.
" Avoirs," 37.

Balance ball, when lawful, 46.
Barrel of beer, definition of, 31.
Barrow, used in weighing coals, 47.
Board of Trade, powers and duties of, 15, 19.
——————— standards, 9.
Bottle and half-bottle measures, 10.
Bottles of wine, 35.
Brand, counterfeiting of, in Ireland, 114.
Bread, sale of, 129–133.
Bushel, definition of, 4.
———, heaped, 5.
Butter, weight and packing of, 31.
———, increasing weight of, in Ireland, 115.
Bye-laws, power to local authority to make, 25.

Cambridge, supervision of weights and measures in, 82.
Capacity, imperial measure of, 1.
Case, stating a, 69.
"Cased" weights, 48.
"Cental" weight, 11, 39, 50.
Certificate of stamping, 57, 58.

Chain, definition of, 1.
Chaldron, definition of, 4.
Cheating, selling by false weights is, 41.
Cheese, sale of, 30.
Clerks of markets to inspect goods sold, 88.
Coals, sale of, 117-129.
Coal measure, 30.
Coal Mines Regulation Act, 91, 92.
Coin weights, definition of, 84.
—————, verification of, 19.
—————, stamping of, 52, 53.
Coke, sale of, 4.
Combination of local authorities, 21.
Commitment, when and how issued, 69.
Contracts to be in terms of imperial weights and measures, 31.
—————, how to be made in Ireland, 101.
Conviction, form of, 68.
Costs may be awarded, 67.
—— of appeal, 77.
Counterfeiting of brand in Ireland, 114.
Court of Summary Jurisdiction, 83.
Custody of standards, 16, 18.
Customary measures not lawful, 32, 34.

Definition of terms, 83.
——————— in Scotland, 97.
——————— in Ireland, 109.
Denomination of weight to be stamped on top, 48.
Diamonds to be sold by troy weight, 32.
Distress warrant, when and how issued, 68.
Dram, definition of, 3.
Drugs may be sold by retail by apothecaries' weight, 32.
Dublin, local authority in, 103.

Earthenware jugs and drinking cups, 50, 51.
Enforcing attendance of defendant, 66.
————————— witnesses, 68.
Entry of appeal, 76.
Error, amount of, allowed in comparison of local standards, 25.
Ex-officio inspectors in Ireland, 105, 106.
Examiners of weights and measures, 26, 27.
Excise penalties, 88, 89.
Expenses of local authority, 20.

Fancy bread, 130, 131.
Fees for stamping weights and measures, 57, 58, 60.
Fiars prices, 93, 94.
Fines, recovery of, 64.
Fish, assize of barrels for, 30.
Fleeces, increasing weight of, in Ireland, 115.
Fluid measures, 10.

Index. 203

Foot, definition of, 1.
Forfeiture of scales, weights and measures, 44.
Forgery of stamp upon weights and measures, 54.
Founders' company, rights of, 81.
Fraud in use of weights, &c., 41, 44.
French bread, 130, 131.
Fringe to be sold by troy weight, 32.
Furlong, definition of, 1.

Gallon, definition of, 4.
Galway, constabulary districts of county of, 103.
Gas-holders, models of, 80.
Glasses of beer and wine, sale by, 35.
Gold to be sold by troy weight, 32.
Grain, definition of, 3.
Gauging of wine casks, 30.

Habere pondus, 37.
Handsale weight, 30.
Heaped measure, 4, 5.
——————— not lawful, 32, 34.
Hearing of appeal, 76.
————— summonses, 66, 67.
Herrings, assize of casks for, 30.
Hindering an inspector, 61, 62.
Hobbit, sale by, 33.
Honey, sale by measure of, 30.
Hundredweight, definition of, 3.

Imperial measures of length, weight and capacity, 1.
Inch, definition of, 1.
Informer may give evidence, 67.
————— may have half the penalty, 70, 71.
Information, how and when to be laid, 65.
Inquisition for rents and tolls payable, 86.
Inspectors, how appointed, 55.
——————, misconduct of, 63.
——————, duties of, 55.
——————, actions against, 78.
Ireland, 101-116.

Justice of the Peace may enter shops, 60.

Lace, to be sold by troy weight, 32.
Lead or pewter weights, 48, 51.
Leet jury, power of, 81, 82.
Legal proceedings, 64-80.
Length, imperial measures of, 1.
Libra civilis, 37.
Libra medica, 37.
Libra et uncia Trojana, 37.

Link, a legal measure, 2.
Local administration, 19.
Local authority to appoint inspectors, 56.
Local comparison of standards, 23.
Local measures not lawful, 32, 34.
Local standards, 11, 23, 24.
London, saving as to, 81, 82.
Long weight, sale by, 36, 37.

Magna Charta, provisions of, 29.
Markets, owners of, to provide scales, 87, 90, 91.
Materials of forfeited weights and measures to be sold, 70, 71.
Measures, imperial, 1.
Metal and glass measures, 49, 52.
Metric weights and measures, 13.
Mile, definition of, 1.
Milk, sale of, 35.
Mill, balance and weights to be kept in every, 90.
Mines, weights used in, 91.
Misconduct of inspector, 63.
Moiety of fine may be paid to informer, 70, 71.

Notice of appeal, how given, 73, 75.

Obstructing an inspector, 61, 62.
Offences, prosecution of, 64.
Orders in Council, 80.
Ounce, definition of, 3.
—— troy, definition of, 3.
Oxford, supervision of weights and measures in, 82.

Parliamentary copies of imperial standards, 8.
Peck, definition of, 4.
Pennyweight abolished, 3.
Pennyworths of milk, 35.
Perch, definition of, 1.
Pint, definition of, 4.
Platinum, to be sold by troy weight, 32.
Plug of lead or pewter for stamping, 48, 51.
Pole, definition of, 1.
Police stations, weighing machines for coals to be kept at, 128.
Possession for use for trade, 42.
——————, evidence as to, 71, 72.
Precious metals and stones to be sold by troy weight, 32.
Price list and price current, 39.
Production of local standards, 24.
Prosecution of offences, 64.

Quart, definition of, 4.
Quarter, definition of, 4.
Quarter Sessions, definition of, 83.

Rate, local, 20.
Recognizances upon appeal, 73, 74, 76.
Recovery of fines, 64.
Rents and tolls, inquisition for, 86.
——— ———, in Scotland, 93.
Repeal of Acts, 84.
Report of Board of Trade, 16.
Restoration of standards, 7-9.
Rod, definition of, 1.
Rood of land, definition of, 2.

Sack as a measure, 4.
Sale of false weights, 42.
Sales of Bullion Act, 3.
Sale of salt, 31.
Savings and definitions, 80.
Schedules, effect of, 80.
Scotland, 93-100.
Scruple, definition of, 4.
Second offences, limitation as to, 71, 72.
Secondary or Board of Trade standards, 9.
Sessions, definition of Quarter, 83.
Sheriff or sheriff substitutes, power of, in Scotland, 99.
Silver to be sold by troy weight, 32.
Small Penalties Act, 1865, 69.
Soke of Peterborough to be a county, 22.
Special case from Quarter Sessions to Queen's Bench, 77.
——————— to a superior Court of Law, 78.
Stamping, definition of, 50, 83.
——— weights and measures, 48, 57.
Standard gallon, imperial, 4.
——— pound, imperial, 2.
——— yard, imperial, 1.
Standards of measure and weight, 7.
Stating a "case," 69.
Stone, definition of, 3.
"Striking" a measure, 6.
Sub-standards in Ireland, 102, 103.
Summary Jurisdiction Act, 64, 83.
——— proceedings, 64, 70.
Summonses, number of, 45.
——————, how served, 65.
——————, hearing of, 66, 67.

Ton, definition of, 3.
Trade, definition of, 14, 31.
Troy weight, 3.
——— ———, origin of, 37.
Tumbrill for testing coin, 29.

Unauthorised weights or measures, 32.
Unit of capacity, 4.
—— weight, 3.
Unjust weights and measures, 41.
Use of weights and measures, 29.

Verification of standards, 16, 17, 18, 23.
———————— weights and measures, 48, 57.
Vessels not containing imperial measure, 32, 39.
Vestry in metropolis in certain cases may appoint inspectors, 26, 27, 28.
Vinegar, barrel of, defined, 31.

Warden of the standards, 15.
Warrant, how and when issued, 65, 66.
Water measure, 31.
Weigh-masters in Ireland, 113.
Weight, imperial measures of, 1.
Weights and measures, use of, 29.
Weighing machine, what is a, 43.
Weighing, mode of, in Ireland, 102.
Westminster, saving as to city of, 81, 82.
Winchester bushel, 30.
Witnesses may be summoned to attend, 68.
Wooden or wicker measures, 50.

Yard, imperial standard, 1, 2.

Recent Works.

LONDON:

KNIGHT & COMPANY.

Index to Subjects.

Subject	PAGE
Artizans' Dwellings	12
Assessment	24
,, Metropolis (See also Rating)	17
Ballot Act	18, 21
Bankruptcy	16
Boarding-out	17
Book-keeping for Public Institutions	16
Burial	24
Bye-Laws	6, 7
Canal Boats	16
Cattle Diseases	19
Chairman's Handbook	8
Churchwardens	15
Clergy	13
Criminal Code	20
Diary	26
Directory	27
Education Law	9, 12
Elections — Parliamentary and Municipal	18, 21
,, School Board	17
,, Guardians	18
,, Local Boards	17
Explosives	20
Highways	6
Infectious Diseases	23
Inspector of Nuisances	18
Licensing	10
Local Government	5
,, Directory	27
,, Orders	6
,, Reports	28
Lunacy	14
Master and Matron of Workhouse	22
Medical Officers of Health	18
,, Poor Law	22
Mortgage Registers	21
Nuisances	12, 18
Official Publications :—	
Bankruptcy	29
Cab Fares	31
County Courts	29
Divorce	30
Education Department	29

Subject	PAGE
Official Publications—*continued*.	
Friendly Societies	29
Land Registry	29
Local Government Board	28
Probate	30
Rules and Orders—Law Courts	29-31
Shipping	29, 31
Orders of Local Government	6
Overseers	6
Parochial Cash Book	13
Paupers	22
,, Education	19
Poor Law Accounts	19, 22
,, Amendment Act	19, 24
,, of Elberfield	24
,, in Belgium	24
,, Conferences	21
,, Orders	6
Poor Removal	21
Prisons	5
Public Health	5, 10
Rating	15, 22
,, of Machinery (See also Assessment)	13
Registration of Voters	11
,, Births and Deaths	20
Relief	17
Rivers Pollution	17
Sanitary Law	6
,, Hints	12
,, Accounts	19
School Board Law	9, 12
,, Accounts	20
Sewage	10, 29
Tables :—	
Acreage, &c.	25
Dates	26
Loans	26
Provisions	25
Rates and Taxes	25
Tenements	26
Thames Valley Sewerage	10
Town Clerk	18
Trade Marks	7, 32
Vaccination	11, 24
Weights and Measures	10
Workhouses	16

KNIGHT & CO.'S complete Catalogue of Books and Forms for the use of Local Authorities in England and Wales may be had on application. It contains Lists of Forms for Poor Law Unions, Urban and Rural Sanitary Authorities, Overseers and other Parish Officers, Highway Boards, School Boards, School Attendance Committees, Registrars, Returning Officers (Parliamentary, Municipal, School Board, &c.), Bankruptcy Courts, Registers and Certificates for the Clergy, Burial Boards, Asylums, &c.

The following may also be obtained: I.—a new and extended Catalogue of over 1000 Blank Precedents. II.—a List of Official Books and Forms issued by Authority for Her Majesty's Stationery Office, which includes New Forms under the Factory and Workshop Act, 1878, the Rules and Orders of the County Courts and other Courts of Law, &c.

90 Fleet Street,
February 1879.

RECENT WORKS.

The Law of Public Health & Local Government, Ninth Edition.

Embracing the Public Health Consolidation Act, 1875, and all other Acts giving Powers to Local Authorities. By W. CUNNINGHAM GLEN, late Principal of the Legal Department of the Local Government Board, and ALEX. GLEN, M.A., LL.B., Barristers-at-Law.

Price 36s. Cloth, 42s. Law Calf.

Law Times.—"There is no book on the subject bearing any comparison with this one by Mr. Glen. .... The work must be considered absolutely complete."

Public Health.—"We cannot speak in too high terms of Mr. Glen's 'Law of Public Health.' It affords throughout evidence of the care and accuracy with which it has been compiled, and which, with the vast amount of information given, renders it an indispensable guide. An index, extending over eighty pages, facilitates reference to any part of this *magnum opus.*"

DEDICATED BY PERMISSION TO THE SECRETARY OF STATE FOR THE HOME DEPARTMENT.

The Law of Prisons in England and Wales:

Containing an Analysis of the Law, together with the Statutes now in force relating to Prisons, including the Prisons Act, 1877; and the New Prison Rules, 1878. By ROBERT WILKINSON, M.A., Cantab., of Lincoln's Inn, Barrister-at-Law.   Price 6s.

Law Times.—"An examination of this volume has satisfied us that it has been compiled with great care and discrimination, and it cannot fail to prove a most useful guide on the subject with which it deals."

Saturday Review.—" Mr. Wilkinson has prepared a volume of great completeness, which contains the Prisons Acts of 1865 and 1877, with an analysis of these Acts, and selections from other Acts still in force on the subject, to which are added some valuable notes, and a full index."

Metropolitan.—" Mr. Wilkinson has, in order to make it complete, supplied selections from other Acts, numerous notes, and a full, elaborate index, which will enable any point to be discovered in a moment. The book is very clearly printed, is of a handy size, and will no doubt become a *vade mecum* with all who have to work under the Acts."

KNIGHT & CO., 90 FLEET STREET.

The Highway Acts, 1862—1868. Fifth Edition :

By ALEXANDER GLEN, M.A., LL.B., Barrister-at-Law. This Edition contains the Highways Amendment Act, 1878, and all the important Decisions of the Courts subsequent to the previous editions. To which is added a paper on the Construction and Management of Roads, including Asphalte and Wood Pavements, by Wm. Nethersole, C.E.   Price 8s. 6d.

Practical Magazine.—"Should be in the hands of every solicitor. Is an exposition of the powers and duties of those who have the management of Highways."
Iron.—"In this Edition (the Fourth) many important improvements have been effected."

Local Government Board Orders; Poor Law, Sanitary, Vaccination and Education :

This volume contains the whole of the Orders now in force issued by the Poor Law Commissioners, the Poor Law Board, and the Local Government Board. The Circulars and other Explanatory Documents issued in relation to the Orders, Notes of Reference, and Index. Preceded by a Tabular Statement shewing the Orders in force in each Union and Parish.   Price 12s.

Local Government Chronicle.—"In short the work is thoroughly good and useful, complete and trustworthy."
Metropolitan. "Will be found a guide to the proper administration of the poor laws and all duties carried out by local authorities."

Manual for Overseers, Assistant Overseers, Collectors, & Vestry Clerks. Fourth Edition:

Containing full and plain Instructions as to their Duties and the mode of Keeping their Accounts. By HUGH OWEN, JUN., Barrister-at-Law.   Price 4s. 6d.

Metropolitan.—"A clear and succinct guide as to what to do and how to do it."
Stamford Mercury.—"A very useful work."
Iron.—"Like all Manuals issued by the same firm every care has been taken to make the work a valuable addition to the best books of reference."
Bookseller.—"A useful Manual to parochial officers."

Model Bye-Laws of the Local Government Board:

For the guidance of Sanitary Authorities, together with the Letters of the Board relating thereto.   In one volume price 3s.

(*Sold also separately in 9 parts.*)

KNIGHT & CO., 90 FLEET STREET.

The Law of Bye Laws;

With an Appendix containing the Model Bye Laws issued by the Local Government Board, the Board of Trade, and the Education Department. By the late W. G. LUMLEY, LL.M., Q.C., Counsel to the Local Government Board, and to the Education Department. Price 10s.

Law Times.—"Should be in the hands of all whose duty it is to make and enforce Bye Laws."

Saturday Review.—"The arrangement appears to us peculiarly good . . . No body having the power to frame Bye Laws need be at a loss as to the proper manner of carrying out its task. Mr. Lumley is therefore to be congratulated on having hit upon and ably treated a hitherto almost untouched branch of the law, and one which is also of considerable and daily increasing importance."

Solicitor's Journal.—"This book not only supplies a want which must have been felt in many quarters of late, but it is of no small historical and archæological value, and is moreover pleasantly written."

The Examiner.—"Contains a good deal of interesting archæology, and no inconsiderable amount of valuable information."

The Economist.—"Mr. Lumley's work is all that could be desired. He has spared no labour to make it complete."

Scotsman.—"The work will be hailed with gratitude."

School Board Chronicle.—"Local authorities generally will find it desirable to have this book in their offices."

Local Government Chronicle.—"Will long remain the standard work upon the subject."

Iron.—"This work will be of great value."

Spectator.—"Primarily for professional readers—but laymen may learn a good deal from its pages."

The Law of Trade Marks:

CONTAINS—Preface and Introduction — Of the Nature of a Trade Mark—What will, and what will not, constitute a Trade Mark—Of Infringement of Trade Marks—Of Fraud in respect of, or connected with, Trade Marks—The Trade Marks Registration Act, 1875—The Trade Marks Registration Amendment Act, 1876—The Trade Marks Registration Extension Act, 1877—Order in Council, 12th December, 1877—Rules under the Trade Marks Registration Acts, 1875-6—Instructions to Persons applying for the Registration of Trade Marks—Index. By C. STEWART-DREWRY, of the Inner Temple, Barrister-at-Law.   Price 3s. 6d.

Metropolitan.—"A valuable treatise on the subject, which is unusually interesting."

BY AUTHORITY. PUBLISHED FORTNIGHTLY.

Trade Marks Journal:

Containing an Engraving of every Registered Trade Mark—Name and Address of Owner—Description of Goods, &c. Parts 1 to 150 are published.   Price 1s. each.

Index to the above—May to December, 1876; January to June, 1877; June to December, 1877; January to June, 1878.
   Price 3s. each.

KNIGHT & CO., 90 FLEET STREET.

RECENT WORKS.

The Chairman's Handbook. Third Edition:

Suggestions and Rules for the conduct of Chairmen of Public and other Meetings, with an Introductory Letter addressed to the Right Honourable the Speaker. By REGINALD F. D. PALGRAVE, the Clerk Assistant of the House of Commons.   Price 1s. 6d.

The Times.—". . . Mr. Palgrave has tried to make the ways of public meetings easy, and the burden of chairmanship light. His position in the House of Commons gives him authority' The result is a most useful and timely Handbook. No question which may reasonably be expected to arise from the mode of electing a chairman down to the happy time at which the business is settled and the meeting adjourned, is left without a simple solution. . . ."

Morning Post.—" It would be impossible to over-estimate the value of the Chairman's Handbook."

Pall Mall Gazette.—" Many useful suggestions."

Morning Advertiser.—" This convenient handy book, a code of clear and lucid rules."

Daily News.—" We agree with Mr. Palgrave that the standard to which Chairmen should seek to conform is that embodied in the well-tried practice of the House."

Daily Telegraph.—" A valuable aid not only to Chairmen, but to all who take part in public meetings."

Newcastle Daily Chronicle.—" We must again indicate our high sense of the value of this little work."

Scotsman.—" We very much doubt whether there could arise any question at a public meeting which a reference to his little volume would not at once determine."

Law Times.—" Cannot fail to be of service."

Saturday Review.—" Put together in a clear and concise form."

Athenæum.—" The rules laid down will be found most useful."

Second Notice.—" Has met with the success it deserved. The author has carefully revised it."

The Guardian.—" An invaluable assistant and undoubted authority."

Record. —" Embodies the usuage and practice of the House of Commons in all that relates to debate, votes, amendments, adjournments, &c."

Rock.—"As an absolute authority upon all matters relating to the conduct of public meetings, this should be carefully studied by all who are in the habit of taking part insuch gatherings."

Economist.—" This is a handy little book. Nobody is better qualified than the author to speak with authority upon the matter."

Graphic.—" This useful little manual."

Illustrated News.—" Of great service to all who take part in public meetings."

Banker's Magazine.—" Exactly the book which every Chairman will be glad to have in his pocket."

Nonconformist.—" The reader knows the subject from the title, but cannot know, without the work, with what practical sagacity and experience Mr. Palgrave has treated it."

Builder.—" It will prove a useful help. The fact that it is written by Mr. Palgrave is an assurance that it may be depended upon."

Architect.—" Should be at the side of every Chairman desirous to prevent irregularity."

Iron.—" Gives many hints on the subject, which will be of general utility."

Leeds Mercury. —"A very useful book. Possessing an authoritative guide like this Chairmen should find no difficulties."

Metropolitan.—" With good effect a copy might well be placed on the table at every public meeting."

Stamford Mercury.—" Exceedingly useful, not only to novices in debate, but also to Chairmen of experience."

Eastern Morning News.—" It will be a great help to many."

School Board Chronicle.—" It affords a great deal of excellent information, much good advice, and a body of rules well worthy of attention."

Bookseller.—" A well-compiled and valuable handbook for public men and speakers."

Local Government Chronicle.—" The information is full, precise, aud adequate; the rules distinct and comprehensive; the directions clear."

Elgin Courant.—" This little book is of much value."

Sheffield Post.—" Mr. Palgrave has done his work very well—an admirable guide.'

KNIGHT & CO., 90 FLEET STREET.

RECENT WORKS.

The Education Acts Manual. Fourteenth Edition:

Embracing the Education Acts, 1870, 1873, 1874, and 1876.
By HUGH OWEN, Jun., Barrister-at-Law.   Price 12s. 6d.

School Board Chronicle.—"The constant companion of all School Board Clerks, Chairmen, and Members of Boards who pride themselves upon being well 'posted up' in the duties in which they are engaged."
    Second Notice.—"Crammed with information, always to the point, and always correct. If the work were not already in existence it would not now be possible to produce such a book."
    Third Notice.—"It seems to us indispensable."
    Fourth Notice.—"More perfect and more useful."
    Fifth Notice.—"Owen's Manual has become a necessary part of the machinery of the School Board system."
    Sixth Notice.—"Owen's Manual is not only a safe guide, but a guide which will very rarely fail whenever happens a difficulty or dead-lock."
    Notice of the 13th Edition.—" . . . . Our readers know well our opinion of the supreme excellence of Owen's Manual. Neither School Board nor School Attendance Committee can dispense with it."
The Standard.—"Well arranged, and full of the most instructive information."
    Second Notice.—"Exceedingly well carried out."
Daily News.—"A well-arranged, clear, and comprehensive exposition."
Hour.—"A guarantee against every sort of irregularity."
Law Times.—"Thoroughly deserves the large measure of popular favour which it has received."
Builder.—"The accuracy and utility of the work are testified by the number of editions through which it has already passed."
Western Daily Press.—"Clear and intelligible . . . A complete *vade mecum.*"
    Second Notice.—"One of the most useful manuals we have seen."
Churchman.—"It is very comprehensive."
The Tablet.—"One of the most perfect and authoritative of its kind."
The Record.—"An intelligible compendium."
The Rock.—"A valuable and well-timed exposition."
    Second Notice.—"All that anyone can need to know respecting the Education Acts is set forth in a clear and comprehensive manner."
Nonconformist.—"Clear, compact, and exhaustive."
Local Government Chronicle.—"The edition now before us will be found as indispensable as its predecessors."
Metropolitan.—"Mr. Owen has smoothed away difficulties."
    Second Notice.—"The fact that this work has reached its eighth edition shows how much the valuable labours of Mr. Owen are appreciated."
Windsor Express.—"Possesses a degree of authority which no other work can obtain."
Education League Paper.—"It is the fullest, and most complete Hand-book."
The Bookseller.—"Will be found of value by every Board."
Examiner.—"Deserving of a favourable recognition."
Carnarvon Herald.—"We advise all interested in Education to possess it."
Echo.—"This manual is very complete."
Leeds Mercury.—"Carefully prepared."
Manchester Guardian.—"An excellent work."
South Wales Press.—"Very useful to all interested in the question."
City Press.—"A well-timed and useful book."
Cardigan Advertiser.—"Of immense value to all interested in working the Act."
Parochial Critic.—"May be read with advantage."
Northern Daily Express.—"Will be found very useful."
Bristol Daily Post.—"A most complete and useful manual."
North Wales Chronicle.—"There must be very few difficulties which are not solved in its pages."
Iron.—"A valuable and trustworthy manual."
    Second Notice.—"The most perfect and authoritative work of its kind."
Eastern Morning News.—"By an author who is now perhaps the greatest authority upon the subject."

KNIGHT & CO., 90 FLEET STREET.

Weights and Measures Act, 1878:

A Card setting forth concisely and clearly the several Offences, Penalties, &c. Price 6d. or 5s. per dozen.

Metropolitan.—"Every one engaged in business should possess a copy."

The Licensing Laws:

The five Licensing Acts are arranged so as to form one comprehensive statute, with copious explanatory notes; together with an introduction giving a full description of the Licensing System, and an appendix of statutes; the whole forming a complete handbook. By GEORGE C. WHITELEY, M.A., Cantab., Barrister-at-Law.
Price 5s.

Law Times.—"A decidedly creditable and useful publication."
Cambridge Chronicle.—"A complete treatise upon the Licensing Laws which it simplifies and explains."
Manchester Guardian.—"Very carefully done . . . especially valuable."
Morning Advertiser.—"The arrangement is excellent, and the value of Mr. Whiteley's opinions cannot be over-estimated."
The Law.—"Mr. Whiteley's work will be most serviceable."
Standard.—"This is a perfect and exhaustive treatise on the whole of the Licensing Laws."

NEW EDITION. BY AUTHORITY.

Sewerage, Drainage, and Water Supply:

Suggestions as to, with Eighteen Coloured Lithographic Plates. By ROBERT RAWLINSON, C.B., Chief Engineering Inspector to the Local Government Board.
Price 3s. paper covers, 4s. 6d. half-bound.

PUBLISHED BY AUTHORITY, AND FOR THE BOARD.

The Law relating to the Lower Thames Valley Main Sewerage Board:

With a Coloured Map of the District. Arranged with Notes, Index, &c., by ALEX. GLEN, M.A., LL.B. Price 5s.

Cheap Edition of the Public Health Statutes:

Containing the Public Health Consolidated Act, 1875; the Rivers Pollution Act, 1876; and the Public Health (Water) Act, 1878; with an Index by F. STRATTON, Solicitor. A well-printed portable volume, bound in limp cloth. Price 3s. 6d.

Metropolitan.—"This is a successful attempt on the part of the Publishers to produce at a moderate price and handy in size, the statutes above mentioned. The idea is a happy one. It is the most compact form in which we have seen the very lengthy Public Health Act, and the book is easy for reference, preceded by a comprehensive index."

KNIGHT & CO., 90 FLEET STREET.

RECENT WORKS.

The Law of Parliamentary and Municipal Registration. Second Edition:

Embracing the Borough Registration Act, 1878, &c.; with an Introduction, Notes, Useful Tables, the most Important Decisions given in Appeal Cases from the Revision Courts, and a full Index. By ALEX. CHARLES NICOLL and ARTHUR JOHN FLAXMAN, of the Middle Temple, Barristers-at-Law. Price 8s. 6d.

From A. G. Marten, Esq., Q.C., M.P.—"A very excellent work; reflecting great credit on the Editors."
From Francis H. Bacon, Esq., of Lincoln's Inn, Revising Barrister.—"A very useful work, and one which was very much wanted."
Law Times.—"A singularly compendious and well-arranged work."
Law Journal.—"Mr. Nicoll and Mr. Flaxman have accomplished their task in a manner that does them very great credit."
Solicitor's Journal.—"The plan of the book appears to be well conceived."
City Press. "A second edition of this valuable work being called for, the authors have made some improvements in it, which cannot fail to render it more generally acceptable."
Globe.—"A very useful book."
Westminster Review.—"A serviceable edition of the Law of Registration."
Fun.—"Another of those books without which every gentleman's library, no matter how otherwise well stocked, must mourn in melancholy incompleteness."
Yorkshire Post.—"A book like this possesses great value."
The Tatler.—"A great benefit to the Legal Profession and the community at large."
Sheffield Independent.—"States in a clear manner the Law relating to Registration."
Lynn Advertiser.—"A useful manual. The editors have been eminently successful in making the law concise and clear."
Oldham Express.—"This book will be of great use."
Bromsgrove Messenger.—"An excellent epitome."
Oxford and Cambridge Journal.—"A treatise upon which greater care could not have been bestowed."
Law Magazine.—"The law is well and concisely set out."
Metropolitan.—"A work which must of necessity be a popular book. It is easy of comprehension, and should prevent much misunderstanding which often arises."

The Law relating to Vaccination. Sixth Edition:

Comprising the Vaccination Acts, the Instructional Circulars, Orders, and Regulations issued by Authority. With Introduction, Notes, and Index. By DANBY P. FRY, Barrister-at-Law, Legal Adviser to the Local Government Board. Price 5s.

Sanitary Record.—"This handy volume contains a complete *vade mecum* for the use of all who are in any way concerned in the administration of the Vaccination Acts. The present edition, like all Mr. Fry's works, is an admirable digest."
The Law.—"Of great practical utility; a reliable guide, brought down to the present time and the most recent decisions carefully noted. It is of a handy size, and well got up in every respect."
Builder.—"The Book is a useful one."
Standard.—"A very compendious and useful Manual."
City Press.—"The Law is clearly stated."
Birmingham Morning Post.—"Of inestimable service."
Bristol Mercury.—"Its usefulness as a guide must be obvious."
Public Health.—"A thoroughly complete handbook."
Metropolitan.—"The best Manual extant on the subject."

KNIGHT & CO., 90 FLEET STREET.

RECENT WORKS.

The Education Act, 1876. Fourth Edition:

With an Introduction, Annotations, and Index, and an Appendix containing the provisions of the Factory Acts as to the Attendance of Children at School, Regulations of the Education Department, and Specimen Bye Laws. By HUGH OWEN, Jun., Barrister-at-Law, author of "The Education Acts Manual," Fourteenth Edition.
Price .

Pall Mall Gazette.—"An exceedingly useful and convenient edition. It can hardly fail to be of great service to those upon whom the new Act has imposed so many new and important duties in which they are engaged."
School Board Chronicle.—"Whatever the difficulties and whatever the pitfalls in the way of those whose business it is to consider and to carry into effect the provisions of the new Education Act, everybody may feel himself perfectly safe in the hands of Mr. Owen He makes no mistakes; he stumbles into no misinterpretations; he never misleads those who rely upon him."
Law Times.—"The Introduction furnishes a digest, and the Notes are of a useful character."
Oxford Times.—"This valuable and inexpensive Book."
Echo.—"Mr. Owen's name, as an Expositor of the first Education Act, is too well known to need a word of commendation."
Literary World.—"Mr. Owen has done well to issue this in a separate form, and has done his work admirably."
Birmingham Post.—Indispensable to persons interested in the working of the Act."
Mercantile Gazette.—"Will doubtless prove of great service to those on whom new and important duties devolve."
Builder.—"The Notes will be of great service to those who will have to administer the Act."
Baptist.—"Mr. Owen's Notes are most useful."
Beehive.—"A most valuable and useful Book."
The Methodist.—"Excellent and well-arranged."

Removal of Nuisances:

Being Practical Hints to Householders, and Officers and Members of Sanitary Authorities. This work supplies useful information as to the law on the subject, written in plain language and adapted either for the professional or general reader.
Price 6d., or 1s. in cloth.

City Press.—"Will be found useful in many emergencies."
Medical Examiner.—"A handy Guide for the Public Health Official whose legal knowledge is frequently and unexpectedly taxed."

Artizans' and Labourers' Dwellings Act, 1 75:

By ALEX. GLEN, M.A., LL.B., Cantab., of the Middle Temple, Barrister-at-Law. With the view of rendering this volume a complete handy book of practical utility, the Introduction to the volume has been written so as to form a summary of the provisions of the Statute; the Index and Table of Contents extend to 25 pages; and at the end is added a collection of Forms required under the Act.
Price 2s. 6d.

Standard.—"A very valuable Guide."
Law Times.—"An extremely useful and handy Guide."

KNIGHT & CO., 90 FLEET STREET.

RECENT WORKS.

A Legal Guide for the Clergy:

With Appendix of recent Statutes, and the Judgment of the Final Court on the Appeal of Mr. Ridsdale. By R. DENNY URLIN, of the Middle Temple, Barrister-at-Law. Price 4s. 6d.

Guardian.—"The contents of this book are clearly set out, and concisely expressed. . . . . In conclusion, we must say, this is a very conveniently arranged and cheap book."

Saturday Review.—"Will save clergymen from the pitfalls offered by ignorance of the law and inability to get information in a compendious form."

Spectator.—" Laymen may learn a good deal from its pages. A concise view of the law."

Literary Churchman.—"For a handy guide to the chief facts, nothing can be better. It incorporates bodily the important ecclesiastical statutes of the last few years."

Law Times.—" A useful and accurate work, suited by reason of its arrangement to the ordinary clergyman."

Oxford University Herald.—" Leaning upon the arm, so to speak, of the author, they need not stumble in the way they have to go."

Birmingham Daily Post.—"The arrangement is admirable, and the articles are written with great clearness and precision. The book is likely to be indispensable for those who have to do with ecclesiastical affairs."

Christian Union.—" Rarely have we met with a work which in so concise a form contained such legal acumen. The classification of the contents of the book is exceedingly commendable. As a work of reference it is invaluable."

Metropolitan.—" We can cordially recommend it. The author deserves the thanks of the clergy for this Book. No churchwarden or sidesman should be without it."

DEDICATED, BY PERMISSION, TO THE LORD BISHOP OF TRURO.

The Parochial Cash Book:

Designed to facilitate the Keeping of Accounts by Clergymen and Churchwardens. This work has been prepared with a view to aiding the adoption of a simple system of Accounts for every parish; it is arranged in a tabular form with printed headings, and shows the following items:—RECEIPTS—Offertories, Voluntary Gifts, Endowments, other Receipts. EXPENDITURE—I. Clergy, Lay-helpers, Church Fabric, Church Services, Day Schools, Sunday Schools, Sick and Poor, Clubs; II. Religious Instruction, Home Missions, Training College; III. Hospitals, Foreign Missions, Sundries. Each Book contains 50 openings, showing the foregoing particulars, and 18 Forms of Annual Balance Sheets.

Price 8s.

(Specimen Sheets sent on receipt of Six Stamps.)

Local Taxation and the Rating of Machinery:

By THOMAS F. HEDLEY. Price 8s.

Railway News.—" The work is an extremely useful one."
Mining Journal.—" Will do much to avoid disputes in future."
Colliery Guardian.—"A book of much value."
Mining World.—"A useful addition to any mining library."
Timber Trades' Journal.—"From it can be learnt all that is necessary as to liability to rating."
Birmingham Gazette.—" Mr. Hedley has placed the great body of local taxpayers under an obligation."

KNIGHT & CO., 90 FLEET STREET.

The Lunacy Laws. Second Edition:

Containing all the Statutes relating to Private Lunatics—Pauper Lunatics—Criminal Lunatics—Commissions of Lunacy—Public and Private Asylums—and the Commissioners in Lunacy; with an Introductory Commentary, Notes to the Statutes, including References to decided Cases; and a Copious Index of 129 pages. By DANBY P. FRY, Barrister-at-Law, Legal Adviser to the Local Government Board. Price 21s.

Saturday Review.—" Mr. Fry has rendered a useful service in putting these Laws together in a connected and intelligible form."

Law Times.—" It is a decidedly valuable compendium of the law of which it treats, which, unhappily, is one of growing importance."

Former Notice.—"There are no books so useful as those which collect all the law on a single subject, arranged so that any part of it may be readily found. This has been the design of the volume before us ; it has been accomplished with care and industry."

Lancet.—"The second edition of this valuable work appears opportunely. It is ably edited, conveniently arranged, and will be indispensable."

Law Magazine.—"The profession is indebted to Mr. Fry for placing at its disposal a work which will be found very useful in the administration of an important department of the law."

Medical Times and Gazette.—" Mr. Danby P. Fry, a barrister of long standing and repute, has given the subject much patient search and labour, and we consider medical men owe him a debt of gratitude for the service he has done."

The Medical Mirror.—"And it is so clearly written that no one having this manual by him need ever be at a loss to promptly ascertain the exact condition of the law upon any particular point concerning persons of unsound mind."

Metropolitan.—" It will undoubtedly be considered a leading authority in Lunacy matters. Its value cannot be too highly estimated, and Mr. Fry must be congratulated upon the success with which he has produced this, the second edition of his work."

Iron.—"The name of the author will be a sufficient guarantee for the completeness of his production."

Public Health.—" It is almost needless to say that it is the standard work of its kind, and that the new edition, with its corrections and additions, will be heartily welcomed. The purpose of Mr. Fry's book has been to render assistance to all charged with the care of insane persons by thoroughly acquainting them with their responsibilities."

Journal of Mental Science.—" This work is invaluable to all connected with the care and treatment of the insane, and it must find a place on the board room table of every asylum in England and Wales. A copious index completes the book, and adds materially to its value."

Carnarvon and Denbigh Herald.—" Thsi is not the first time for Mr. Fry to appear as a legal writer and annotator, and from a very careful perusal of the introductory commentary of the work we now review, and a close inspection of the arrangement of the whole, we have no hesitation in pronouncing it to be a most excellent and serviceable manual."

North Wales Chronicle.—" Mr. Fry's work cannot but be highly serviceable. From the completeness of its matter, its valuable foot notes, its reference to decided cases, its minute index, and elaborate and clear introductory Commentary, is, we are persuaded, well calculated to communicate every information required on the subject."

Inquirer.—" This book is one of the best of its kind that we have ever seen, and mus form a part of the library of every public man. A book of this kind is a necessity, and the author who does such work as this well, renders a service to society. There is not about this work any of that hasty book-making which characterises some of the editors of our Acts of Parliament."

KNIGHT & CO., 90 FLEET STREET.

The Churchwarden's Guide. Ninth Edition:

Edited by W. G. BROOKE, M.A., Barrister-at-Law. This work is not only a Guide to Churchwardens, but forms also a complete Legal Text Book, with Tables of Statutes and Cases brought down to the present time, and including the Public Worship Regulation Act, &c. Price 4s. 6d.

Saturday Review.—"The fact that it has gone through nine editions shows how much such a work is wanted by those who fill that anxious office."
Law Times.—"An invaluable, and, we believe, trustworthy Guide."
Second Notice.—"It states the law with accuracy and clearness."
Churchman's Magazine.—"Drawn up with great care and clearness."
The Rock.—"A most valuable work."
Second Notice.—"The latest changes are duly noted."
Church Times.—"Exceedingly well compiled. A clear and authoritative statement of the duties and liabilities of Churchwardens."
Cambridge Chronicle and University Journal.—"Has been carefully revised, and such alterations and additions introduced as recent legislation or decisions rendered necessary."
Iron.—"A handy compendium of all matters affecting the duties and privileges of Churchwardens."
Stamford Mercury.—"The utility of this Guide is evident."
Norwich Mercury.—"Possessing this book Churchwardens need never remain in doubt as to what would be the proper course to pursue under any circumstances that may arise."
School Board Chronicle.—"The result of all experience and all legal decisions on the subject."
The Metropolitan.—"The whole matter is clearly set forth in a very plain and unmistakable manner."
Builder.—"There are a good many rocks upon which Churchwardens may strike . . . they will find the price of this Guide a good investment."
Sheffield Independent.—"An authentic and full digest, well got up."
Western Morning News.—"It is superfluous to say anything in praise of the Churchwarden's Guide, which has long been so highly valued throughout the country."
Bookseller.—"This new edition seems thoroughly well timed."
Warrington Guardian.—"It seems to arm the Churchwarden at every point."
Sussex Daily News. "We have rarely seen a work that bore such evident traces of careful compilation—it is simply indispensable to every Churchwarden."
Newcastle Daily Chronicle.—"Of considerable use—carefully brought down to the latest date."

The Rating Act, 1874. Second Edition:

With Introduction, Notes, and Index. By DANBY P. FRY, Barrister-at-Law, Legal Adviser to the Local Government Board.
Price 2s.

Morning Advertiser.—"This is a very useful edition, and gives practical information. Mr. Fry is one of the assistant secretaries of the Local Government Board, and is therefore well versed in the question of rating; and his introduction is a clear exposition. The notes give useful practical directions."
Land and Water.—"Those who may wish for fuller information are referred to a carefully written publication by Mr. Danby P. Fry."
The Globe.—"The matter is fully discussed in an interesting little volume by Mr. D. P. Fry, from which we recently drew some facts,"
Field.—"This useful little Work is very serviceable for reference, which is rendered easy by a copious Index."
Metropolitan.—"The whole matter is detailed with notes in this publication."

RECENT WORKS.

The Law as to Canal Boats used as Dwellings:

Containing the Canal Boats Act, 1877, and the Regulations and other Documents of the Local Government Board, &c., with an Introduction, Index, and Notes. By JAMES B. HUTCHINS.

Price 2s. 6d.

Canal Boats Act, 1877:

Regulations of Local Government Board under the—
Price 6d. each, 4s. per dozen.

Book-Keeping for Public Institutions:

Applicable to Asylums, Prisons, Hospitals, Schools, Dispensaries, Societies, Clubs, and other Establishments. With Suggestions in regard to the Supervision of Stores, and Audit of Accounts by Finance Committees, Guardians, and Governing Bodies generally. It explains, in plain language, and by means of brief exemplifications, the mode in which the accounts of such establishments should be kept; and it contains a number of valuable suggestions hitherto unpublished. To Clerks and book-keepers it will be found of great assistance, as well as to persons seeking to qualify themselves for clerical situations in public institutions. By JAMES WILLIAM PALMER, Clerk to the Middlesex County Lunatic Asylum, Hanwell.
Price 5s.

City Press.—"Treated in a clear and intelligible style."
Charity Organization Reporter.—"Contains all that is necessary for Charities and other Public Institutions."
Metropolitan.—"Mr. Palmer has had great experience in the affairs of large institutions, and he knows that many of them, particularly those which have outgrown their original magnitude, do not possess an intelligible yet simple system of keeping their accounts. The work before us sketches out a plan, with illustrations, by which boards of management or committees may acquire all the knowledge requisite to enable them to exercise a proper supervision over the expenditure."

A Guide to the Construction and Management of Workhouses:

By the late EDWARD SMITH, M.D., LL.B., F.R.S., Local Government Board Inspector.
Price 8s. 6d.

The Law of Bankruptcy:

The Volume contains, in addition to the Statutes, the whole of the General Rules and Orders of the Court of Bankruptcy; together with the Prescribed Forms, Scale of Fees, Stamp Duties, &c. By C. W. LOVESY, Barrister-at-Law.
Price 7s. 6d.

KNIGHT & CO., 90 FLEET STREET.

The Law as to the Pollution of Rivers:

Embracing The Rivers' Pollution Prevention Act, 1876. By ALEX. GLEN, M.A., LL.B., Cantab., Barrister-at-Law.

Price 2s. 6d.

Saturday Review.—"Mr. Glen has produced a compact analysis of the law as to the pollution of rivers, showing the new powers given to authorities, and to private persons."
Public Health.—"Messrs. Knight and Co. could not have brought out a more useful little manual. The references to other enactments are very useful, and the decisions of the courts will throw light upon obscure parts. The editor has carefully noted the distinctive sources of river pollution. The notes show a large experience and reading."
Manchester Guardian.—"Local authorities have an excellent guide to their duties in Mr. Glen's compendious text-book."
Iron.—"The Legal Manuals published by Knight and Co. are in good repute, and the present work is no exception as respects fulness and accuracy."
Metropolitan.—"The introduction forms a useful digest, and very necessary notes are appended."

The Boarding-Out System:

A practical Guide to the Boarding-Out System for Pauper Children. Containing Suggestions, Forms, and Regulations. By Colonel C. W. GRANT, R.E., J.P. for Somerset, &c. Price 1s. 6d.

Pall Mall Gazette.—"Colonel Grant's little handbook may be consulted with profit."

The Valuation (Metropolis) Act, 1869. Second Edition:

With Introduction, Explanatory Notes, &c. By DANBY P. FRY, Barrister-at-Law, Legal Adviser to the Local Government Board.

Price 3s. 6d.

Parochial Critic.—"We have rarely met with a Book that so clearly explains an Act of Parliament."

The Election of Local Boards:

And the Constitution of Local Government Districts under the Public Health Act, 1875. By ALEX. GLEN, M.A., LL.B., Barrister-at-Law. Price 3s. 6d.

The School Board Election Manual:

Being a complete guide as to procedure in Elections of Boards, and Applications for Boards, in Boroughs and Parishes. By HUGH OWEN, Jun., Esq., Barrister-at-Law. Price 3s.

School Board Chronicle.—"The instructions are full, clear, and circumstantial."
Bristol Daily Post.—"It cannot fail to be useful."

KNIGHT & CO., 90 FLEET STREET.

Manual for Medical Officers of Health. Second Edition:

By the late EDWARD SMITH, M.D., LL.B., F.R.S., Inspector to the Local Government Board. Price 8s. 6d.

Law Times.—"Will prove a most useful guide to Medical Officers."
The Medical Press and Circular.—"Well calculated to serve the purpose for which it is published. There must be many who will be glad to have brought before them their duties by one in the position of Dr. E. Smith."
Public Health.—"It is written by Dr. Edward Smith, one of the Inspectors of the Local Government Board, and contains, in a concise form, everything that a medical officer of health would require to know. We can highly recommend it."
The Watchman.—"All thoughtful and intelligent persons will find much to interest and profit."
Doctor.—"Seems to us likely to prove very serviceable."
Iron.—". . . . Thus, in the course of 346 pages, Dr. Smith has produced a work perfectly unique in the value and variety of its contents, and in its interest, &c."
The Hour.—"The matter is clearly arranged, and the importance of giving a classified table of contents as well as an alphabetical index has not been overlooked."
Metropolitan.—"We owe a debt of gratitude to the author for such valuable assistance."
Lancet.—"Contains many useful practical hints."

Handbook for Inspectors of Nuisances:

Containing full Information as to Appointment and Duties under the Orders of the Local Government Board and the Public Health and other Acts. With numerous Illustrations of Sanitary Appliances. By the late Dr. EDWARD SMITH. Price 5s.

Metropolitan.—"When we consider the great importance of the office of Inspector of Nuisances, we see more and more how necessary it is the officer in question should not only be active, upright, and intelligent, but educated to think and act correctly for himself, and this handbook of Dr. Smith's is just the very work he requires to attain that object."

A Manual for Elections by Ballot—Municipal and Parliamentary. Second Edition:

By HUGH OWEN, Junr., Barrister-at-Law. The present edition of the Ballot Act Manual contains *The Municipal Elections Act*, and *The Returning Officers (Parliamentary Elections) Act* of 1875, by which some material alterations were made in the previous law. The notes in the first edition have been carefully revised, and considerable additions have been made. Price 3s. 6d.

Echo.—"Extremely good and useful."

Election of Guardians. Second Edition:

A Guide to the Practice at Elections of Guardians of the Poor, containing the Statutory Enactments, the Orders of the Local Government Board, including that of 1877. By DANBY P. FRY, Barrister-at-Law, Legal Adviser to the Local Government Board.
Price 5s.

Town Clerk's Memoranda:

A Tabulated Diary. By RICHD. AUBURY ESSERY, late Town Clerk of Swansea. Price 1s.

KNIGHT & CO., 90 FLEET STREET.

Facts and Fallacies of Pauper Education:

By WALTER R. BROWNE, M.A., late Fellow of Trinity College, Cambridge. Price 4d. or 3s. per dozen.

Metropolitan.—"Every Guardian and every ratepayer in the three kingdoms ought to read this invaluable pamphlet, which contains a world of information in a small compass."

The Cattle Diseases Act, 1878:

Together with the new Orders in Council, preceded by an Index.
Price 2s.

The Magnet.—"A handy volume, well-arranged and clearly printed, with an index which will greatly facilitate reference."

Chamber of Agriculture Journal.—"Reduced to very handy dimensions, Messrs. Knight & Co. have published under the above title, the whole law as to cattle disease. A reprint of the Act itself and the full text of the Privy Council Orders, which have to be construed with it, are thus placed side by side in a volume capable of being carried without trouble in the pocket. To have provided this substitute for a variety of separate, bulky, and cumbrous official documents is an unquestioned boon."

Field.—"A publication, in a convenient form, of the Act of 1878 relating to cattle diseases, with the Orders which have been issued by the Privy Council. The work will be found very useful."

Local Government Chronicle.—"An Act which, with the Orders which have to be construed along with it, is one of considerable complexity, is thus rendered as comprehensible and easy of reference."

The Poor Law Amendment and Divided Parishes Act, 1876, &c.:

With Notes, Appendix, and Index. By EDMUND LUMLEY, B.A., Barrister-at-Law. Price 2s.

Accounts of Sanitary Authorities:

An Exemplification of Accounts for Urban and Rural Sanitary Authorities. By GEORGE GIBSON, District Auditor.
Price 17s. 6d.

Metropolitan.—"Of incalculable service to every Sanitary Authority in the kingdom."

Accounts of Guardians and other Local Authorities, under the Education Act, 1876:

An Exemplification and Guide. By GEORGE GIBSON, District Auditor. The Elementary Education Act, 1876, created new duties for Guardians of the Poor, and other Local Authorities, the cost of executing which being chargeable to two different funds, has, in some instances, given rise to uncertainty as to the manner in which the expenses incurred by the Guardians direct, or presented to them for payment by the School Attendance Committees appointed by them should appear in their accounts. This work has been written in the hope of removing these difficulties. Price 3s. 6d.

KNIGHT & CO., 90 FLEET STREET.

RECENT WORKS.

The Law relating to the Registration of Births, Deaths, and Marriages. Second Edition:

By W. CUNNINGHAM GLEN, and ALEX. GLEN, M.A., LL.B., Barristers-at-Law. It contains all the Statutes relating to the registration of Births, Deaths, and Marriages; Non-Parochial Registers; the duties of Registration Officers; and the Marriage of Dissenters in England. To these are added the whole of the Statutes which cast any collateral duty upon Registration Officers, together with the legal decisions upon each Statute. A very full Index completes the Work. Price 5s.

The Law.—"No amount of trouble seems to have been spared to render the book complete . . . Nothing can be better done to clear the subject than is done by the present work, which is in every way excellently got up."
School Board Chronicle.—"A complete exposition of the law . . . a laboriously edited work, a table of contents of a dozen pages, and an index of twenty-six pages. Should be in every library of reference."
Lancet.—"A second edition, much extended, and greatly improved."

Guide Book to the Explosives Act, 1875:

For the use of Local Authorities and their Officers. By Major MAJENDIE, R.A., H.M. Inspector of Explosives. *By Authority.* Bound in leather, 320 pages. Price 2s.

Criminal Code: The Proposed New

Re-issue in 8vo., *by Authority.* Price 2s. 6d.

An Exemplification and Guide to the Keeping of the Accounts of School Boards:

By GEORGE GIBSON, one of the District Auditors of the Local Government Board. Price 6s.

School Board Chronicle.—"To know how to keep School Board Accounts demands some study and some intelligence; but the thing can be learned by any educated man of average capacity, and Mr. Gibson has, by the publication of this book, made the task easy and safe. School Board Accounts are kept, not according to any general rules, but in strict accordance with the general order of the Local Government Board, which rigidly prescribes the form in which, and the persons by whom, they shall be kept: and Mr. Gibson gives in this volume all the requisite information to the utmost detail, with forms, examples, and everything that is necessary to enable a newly-appointed School Board Clerk to set about the work and to carry it out in due and proper form. It is a book which may be absolutely relied upon."
Metropolitan.—"Messrs. Knight & Co. have just issued a work by Mr. George Gibson, which will prove useful. It is, as described in the title, 'An Exemplification and Guide to the Keeping of the Accounts of School Boards.' Mr. Gibson is the author of a very elaborate, yet intelligible, work of a similar character, for the use of Urban and Rural authorities; and, being a district auditor, is of all men best qualified to explain how School Board Books should be kept. We have seen similar productions by other authors, but we must say that, for clearness and completeness, this is the best. The book is of handy size, and well got up."

KNIGHT & CO., 90 FLEET STREET.

Practical Legislation, or the Composition of Acts of Parliament:

By Sir Henry Thring, K.C.B., Parliamentary Counsel.
Price 2s.

Parliamentary and Municipal Elections by Ballot:

With Notes and Observations on the Defects of the System, including concise Instructions to Returning Officers, etc. By Richard Aubrey Essery, Attorney-at-Law, late Town Clerk of the Borough of Swansea. Price 3s. 6d.

Poor Removal and Union Chargeability Acts. Second Edition:

With an Appendix, containing the Acts relating to the Removal of Scotch and Irish Poor, and the Metropolitan Houseless Poor Act. By the late W. G. Lumley, Q.C., Counsel to the Local Government Board. Price 5s.

Private Bill Legislation. Second Edition. (1859):

Shewing the steps to be taken by Promoters or Opponents of a Private Bill before and after its Presentation to Parliament, and the Standing Orders of both Houses. By S. B. Bristowe, Q.C.
Price 6s.

Reports of Poor Law District Conferences:

In 8vo, bound in cloth, with Index, &c.
The Volume for 1875, price 6s. 6d.
Ditto 1876, price 7s. 6d.
Ditto 1877, price 5s. 6d.
Ditto 1878, price 8s. 0d.

Register of Mortgages and Transfers, and Mortgage Ledger:

For the use of Public Boards. Constructed by William Rees, District Auditor and Consulting Accountant.
Contents:—Register of Mortgages—Register of Transfer of Mortgages—Borrowing Powers—Money Borrowed for—Redemption Fund Account—Sinking Fund Account—Redemption of Mortgages, Annual Statement—Personal Account of Mortgagee.
Price, in green vellum, 25s.

Register of Mortgages:

Another Form, foolscap oblong. Price 7s.

Knight & Co., 90 Fleet Street.

Manuals for Poor Law Officers:

By the late WILLIAM GOLDEN LUMLEY, Q.C., Counsel to the Local Government Board.

The Regulations which relate to Poor Law Officers are found in separate Orders, and in many distinct communications, and are for the most part intermingled with others relating to different Officers and Subjects of an entirely separate character. It has appeared to be convenient to detach for the use of each Officer all that relates to him, both as regards his Appointment and his duties, so that he may readily find, in his own Manual, what it concerns him to know, without having to seek for it in the large collection.

Master and Matron of the Workhouse. Second Edition. 4/6
Relieving Officer. New Edition in preparation
Treasurer 1/6
Medical Officer. Third Edition........................... 6/0

Lancet.—"This is a most useful manual. ... Should be in the hands of every medical officer."

The Poor Law Board's Order of Accounts:

An Exemplification of the General Order of Accounts (dated 14th January, 1867). By DANBY P. FRY, Barrister-at-Law, Legal Adviser to the Local Government Board.

Bound in leather. Price 12s.

The following portions of the Work may be had separately:—

	s.	d.
The Overseers' and Collectors' Accounts	1	6
The Master of the Workhouse do.	3	0
The Relieving Officer do.	1	6

Penfold on Rating. Sixth Edition:

Practical Remarks upon the Principle of Rating Railway, Gas, Water, and other Companies; Land, Tithes, Buildings, Manufactories, and other Properties. Sixth Edition, Re-written and Extended, by J. T. KERSHAW and W. MARSHALL. All the more important decisions of the Court of Queen's Bench on questions of Rating are given, and it shews the manner in which the principles laid down are practically applied. Price 10s. 6d.

Mechanics' Magazine.—"A standard book of reference."
Herepath's Railway Journal.—"Written with remarkable clearness."
Estates Gazette.—"The whole question of Rating is clearly and concisely treated."
Railway Record.—"This work is without a rival.... Clears up many doubtful points."
The Builder.—"We can recommend the volume."
Railway News.—"Justly regarded as a leading authority."
Chamber of Agriculture Journal.—"A standard work for the valuer, land agent, lawyer, member of assessment committee, or possessor of rateable property."
Sheffield Daily Telegraph.—"We can cordially recommend this work."

The Pauper Inmates' Discharge and Regulation Act, 1871. Second Edition:

With Introduction and Notes. By HUGH OWEN, Jun., Barrister-at-Law. Price 1s.

KNIGHT & CO., 90 FLEET STREET.

Precautions against Infectious Diseases, &c.:

Precautions against the infection of Cholera—Circular for Distribution, issued by the Local Government Board, July 5, 1873. 50 copies, 2s. 6d.; 100, 5s.; 250, 10s. 6d.; 500, 18s. 6d.; 1000, 30s.

Precautions against Cholera—Placard for Posting. 100, 6s.; 500, 24s.; 1000, 40s.

Memorandum of Privy Council on Disinfection. Price 1s. 6d. per dozen, or 8s. 6d. per 100.

Hospital Accommodation— Official Memorandum as to accommodation, to be given by Local Authorities, together with a Plan of a Temporary Building for holding 16 patients, eight of each sex.
Price 6d.

Plain Advice to all during the Visitation of the Cholera—
Folio sheet, price per 100, 3s.
The same in the Welsh Language. Price per 100, 4s.
——————————French Language. Price per 100, 4s.
——————————German Language. Price per 100, 4s.

Plain Advice—Printed as a Tract. To which are added Directions and Precautions, addressed to Individuals or Heads of Families, &c.
Price per doz., 1s. 6d.

How to Meet the Cholera. Third Edition—A Guide for Parish Officers and others. By J. J. SCOTT, Esq., Barrister-at-Law. Price 6d.

Precautions against Small Pox—For Boards of Guardians, Local Boards, and other Authorities. By JAMES B. HUTCHINS, of the Medical Department of the Privy Council. Price 6d. each, or 5s. per dozen.

Medical Times.—" No 'Board' can go wrong which follows this Guide."

Precautions against Scarlatina—Placard. Per 1000, 30s.

Plain Directions for Preventing the Spread of Infectious Diseases— such as Small Pox, Scarlatina, Fever, Diphtheria, &c. By Dr. W. N. THURSFIELD.

Price, as hand-bill or placard, per 100, 4s. 0d.
200, 7s. 6d.
500, 16s. 0d.
The same, in pamphlet form. Per dozen, 2s.; per 100, 12s.

Some Plain Instructions for Preventing the Spread of Infectious Diseases —such as—Measles, Small Pox, Fevers, Diphtheria, Measles, &c. By the West Kent Medical Officer of Health.
Per dozen, 2s.; per 100, 12s

Suggestions as to Infectious Diseases, &c.—By the Society of Medical Officers of Health. Price 6s. per 100.

KNIGHT & CO., 90 FLEET STREET.

Suggestions to the Charitable. Third Edition :

Being Suggestions for Systematic Inquiry into the Cases of Applicants for Relief. With an Appendix containing extracts from Statutes, Regulations of the Local Government Board, and other matters relating to the Poor. By C. J. RIBTON TURNER, of the Charity Organization Society. Price 2s. 6d.

The Poor Law System of Elberfeld :

Being a Report to the Local Government Board. By ANDREW DOYLE, Inspector. Price 6d.

Relief of the Poor in Belgium :

An Account of the Administrations and of the Reformatory Schools of Ruysselede and Beernem, with lithographed plan. Edited by F. T. BIRCHAM, Assistant Local Government Board Inspector.
Price 1s.

Poor Rate Assessment and Collection Act. Sixth Edition :

With Introduction. Copious Notes, and Index. By HUGH OWEN, Junr., Barrister-at-Law. Price 2s.

Poor Law Amendment Act, 1868. Second Edition :

With Notes, and an Appendix. By HUGH OWEN, Junr., Barrister-at-Law. Price 1s. 6d.

Scott's Burial Acts, Metropolitan & Provincial, 1852 to 1870. Seventh Edition :

In Preparation.

Vaccination Officer's and Public Vaccinator's Handbook. Second Edition :

By WALTER BULLAR ROSS, Solicitor, Clerk to Guardians, &c.
Price 4s.

Medical Times.—"We can strongly recommend it."
Solicitor's Journal.—"The Vaccination Officer will do well to put himself under Mr. Ross's guidance."
The Medical Record. "Few we imagine, can afford to dispense with this book."
Law Times. "We can honestly recommend it."
Ipswich Journal.—"An Index of no less than 35 pages . . .very neatly got up."
Suffolk Mercury.—" Mr. Ross has done good service to the community."
Suffolk Chronicle.—"An admirable *vade mecum.*"
Metropolitan.—" How to master every detail in connection with the enforcement of Vaccination, is clearly and fully set forth."
School Board Chronicle.—"Seeks to move Local Authorities and Officers to such mode of performing their work as will lead to the desired result."
Bury Post.—"The fullest and most lucid information as to the construction of the law, and may safely be relied on."

KNIGHT & CO., 90 FLEET STREET.

RECENT WORKS.

Complete Rate and Tax Calculator:

Or the Rate and Tax Maker's and Rate and Tax Payer's Companion; for calculating the Income and other Taxes, Poor Rates, Church Rates, Highway Rates, Water Rates, &c., containing upwards of 1,700 Calculations, and showing the value of any sum from 2d. to £1000, at One Farthing to Two Shillings and Sixpence in the Pound; arranged so as to render each sum a collectable amount. By JAMES VALENTINE WARD. Revised and recommended by GEORGE BARNES, District Auditor. Price 5s.

Workhouse Provision Tables:

Sanctioned by the Poor Law Board.
For Calculating the Quantity of Provisions consumed by any number of Persons according to the Workhouse Dietary, and also the loss incurred in Cooking, Cutting, and Weighing Provisions, &c.
Price 1s.

The Ready Reckoner of Acreage and Rental and The Assessment Calculator:

Showing the value per acre when the acreage and rental are known; or the rent, when the acreage and value per acre are given; or the acreage, when the value per acre and rent are given, together with the *Assessment Calculator*, a set of Tables showing at once the Rateable Value, after deducting from 2½ to 20 per cent. on Rentals £1 to 2000. Price 2s. 6d., cloth.

The Rate Reckoner. New Edition:

Showing at a glance the Amount of Deduction at 15, 25, and 30 per cent. respectively, and the Net Amount of Rate payable by Owners for any Amount of Rates, from 1d. to £500. By ALFRED JAMES ROBERTS, Vestry Clerk, Hammersmith. 8vo., cloth.
Price 1s. 6d.

Metropolitan.—"People who don't know how to 'cast accounts,' or who can't find time, had better get this book, which is simplicity itself."

Hills's Sheet Ready Reckoner:

Adapted for the use of Poor Law Officers and General Traders and Dealers. Price 6d.

With this Table, Clerks to Guardians and Relieving Officers can find out any given sum required, from One Farthing to Five Shillings, at a single glance. What is useful for this especial purpose is equally available for the use of Traders and Dealers. There is no Ready Reckoner existing so compact and so cheap.

KNIGHT & CO., 90 FLEET STREET.

RECENT WORKS.

Marshall's Weekly Tenement Assessor:

Containing all the Tables necessary to enable the Gross Rental and Rateable Value of Weekly Tenements to be at once ascertained. By WILLIAM MARSHALL, Surveyor, Railway Valuer, &c. Price 1s.

Tables as to Repayment of Loans:

For the use of Local Authorities, &c., showing the equal annual amount of Repayment of Principal and Interest at 3 to 6 per cent. from 1 to 50 years. Price 1s. 6d.

Tables showing at a glance the Number of Days between any two given Dates within the Year:

By HENRY GEORGE SMITH. Intended chiefly for the assistance of Union Clerks and others in the preparation of the half-yearly Claims for the Government Allowance in aid of the Maintenance of Pauper Lunatics in Asylums. It will also prove useful to Bankers, Solicitors, and others, as a supplement to existing Interest Tables. Price 1s. 6d.

The School Board and Teachers' Directory for 1879:

Price 2s.

The Poor Law Sheet Almanac for 1879:

Price 2s., Plain; 2s. 9d., Varnished; 6s. 6d. on Roller, Varnished.

Knight's Official Diary for 1879:

Size: large 8vo., allowing Half-a-page to each Day. The Diary is carefully ruled and printed on superior paper, and strongly bound, and contains

Notice of Local Government Business to be done each Month—Ambassadors—Appraisement Stamps—Apprenticeship—Assessed Taxes—Assurance Companies—Bankers—Bankruptcy Bills of Exchange Bishops and Deans—Bonds—City of London Court—County Courts, Districts and Judges—Eclipses in 1876—Government Offices—HOUSE OF COMMONS—HOUSE OF PEERS—Income and Property Tax—Insurance Duties—Justice, High Courts of—Leases, Stamp Duties on—Legacy Duties—Licenses—Lords Lieutenant in England and Wales—Ministry—Police Offices—Postal Regulations—Preliminary Notes for 1879—Prince of Wales's Household—Public Offices Quarters of the Year—Queen's Household—Stamps Duties—Terms and Returns—University Terms, &c.

Price 3s. 6d., or 4s. 6d. interleaved with Ruled Paper.

KNIGHT & CO., 90 FLEET STREET.

[38th Year.

The Local Government Directory, 1879:

CONTENTS.   *Price* 8s. 6d.

UNIONS in England and Wales, with the Area and Population of each; the Day of Meeting, Number of Parishes, and of Elected and Ex-officio Guardians, and the Inspectors' and Auditors' Districts; the Names and Addresses of the Clerks and Treasurers: the Names of the Chairmen, Vice-Chairmen, Relieving Officers, Medical Officers, Workhouse Masters, Chaplains, and Teachers, with the Certificates awarded to the latter, and the Situation and Capacity of the Workhouse, &c.

SIMILAR PARTICULARS AS TO

URBAN, RURAL, METROPOLITAN, AND PORT SANITARY AUTHORI-ties. Burial Boards, &c. Particulars of the Local Government Board and its Officers, Auditors, District Schools, Reformatory and Industrial Schools, Public Analysts, Schools and Institutions certified by the Local Government Board, the Lunacy Commission, List of Lunatic Asylums, the Calendar, with Memoranda of Business to be done each Month, &c.

Also an **ABSTRACT OF THE LEGISLATION OF 1878.**

Standard.—"Abounds in a mass of information of great value."
Daily Telegraph.—"Full, concise, and accurate. Recent additions have brought the work to a state of great completeness."
Spectator.—"Messrs. Knight's Directory is of the greatest utility."
Lancet.—"The book is, in fact, a 'requisite' in the office of all clerks."
Medical Press.—"This is a very useful Handbook on Local Government, &c."
The Tablet.—"Full of information arranged in a most convenient form."
The Rock.—"A most useful Work, full of information."
Second Notice.—"This work is one of considerable value, admirably arranged."
Sheffield Independent.—"A mass of details often wanted, but otherwise inaccessible."
Iron.—"Besides the list of public bodies, with the names of their officers, we have an Appendix, giving the titles and prices of the Official Forms, of which Messrs. Knight and Co. are the Publishers."
Second Notice.—"It has grown in extent of interest and usefulness."
School Board Chronicle.—"The mass of information originally compiled here, and revised from year to year . . . must be very convenient."
Sanitary Record.—"Contains a mass of useful information."
Metropolitan.—"One of the most valuable Books of Reference published in England."
Oxford University Herald.—"Considering its real worth it is remarkably cheap."
Parochial Critic.—"This excellent Annual contains a large mass of information."
Builder.—"This is a very useful work."
Second Notice.—"Precisely what it purports to be, and is an absolute necessity for all concerned with Local Government Boards."
Public Health.—"The fact that the firm by which it is produced holds the position of authorized publishers to various Government Departments, is a sufficient guarantee of its correctness."
Watchman.—"This is a very convenient and useful book."
Medical Times.—"Deserves praise and recommendation for its completeness."
Second Notice.—"This admirable Directory."

KNIGHT & CO., 90 FLEET STREET.

Official Publications.

Local Government Board:

	s.	d.
Pauper Children Emigration to Canada, Reply of Mr. Doyle to Miss Rye's Report, addressed to the President of the Local Government Board. Price 6*d*., or per dozen	4	0
Model Bye-Laws issued by the Local Government Board, for the use of Sanitary Authorities, together with the Letters of the Board relating thereto. In One Volume, price (Sold also separately in 9 parts.)	3	0
Reports on Poor Laws in Foreign Countries	2	8
Report on Poor Law Administration in London. By HENRY LONGLEY, Esq., Inspector	2	0
Reports of the Local Government Board :—		
First Annual Report	2	6
Second ,, ...	2	6
Third ,, ...	4	0
Fourth ,, ...	3	2
Fifth ,, ...	4	10
Sixth ,, ...	2	8
Seventh ,, (1877-8)	3	10
Memorandum as to Out-Relief, March, 1878 ... per dozen	2	6.
Proposed District School on the System of Mettray. With a Lithographic View of the Agricultural Colony of Mettray. By ANDREW DOYLE, Esq., Inspector	0	3
Reports of the Medical Officer of the Local Government Board. New Series :—		
Part I. (1874)	2	0
,, II. (1874) with plans	7	6
,, III. (1874)	1	4
,, IV. (1875) with plans	1	0
,, V. (1876-7) with photographs and plans ...	10	6

Insurance and Annuities:

	s.	d.
Regulations, made under 27 & 28 Vict., c. 43, respecting Government Insurances and Annuities	0	6
Table of Premiums for Government Life Insurance and Annuities	0	3½

Settled Estates Act, 1878:

	s.	d.
Orders, December, 1878	0	3

KNIGHT & CO., 90 FLEET STREET.

Sewage Disposal:

	s.	d.
Report of Committee appointed by Local Government Board to inquire into the modes of treating Town Sewage		
Plans to accompany same...		
Drainage—Suggestions as to Sewerage, Drainage, and Water Supply. New Edition. With Plans and Drawings Paper cover, 3s.; half-bound	4	6

Education Department:

	s.	d.
Rules for Planning and Fitting up Schools	1	0
New Code, 1878	0	2½

Bankruptcy:

	s.	d.
General Rules for regulating Practice and Procedure of the London Bankruptcy Court, and of the County Courts, and Scales of Costs, &c.	2	0
Ditto ... large paper	2	6
Further General Rules, July, 1871	0	3
Order regulating Fees to be taken under ditto, August, 1871	0	1
Rules of Court made in pursuance of Bankruptcy Repeal and Insolvent Court Act, 1869	0	2
Ditto ... large paper	0	3

County Courts:

	s.	d.
The Consolidated County Court Orders and Rules, 1875, with Forms and Scales of Costs and Fees	3	0
The County Court Rules, 1876	0	6
Ditto ditto ditto 1877	0	1

Shipping Casualities:

	s.	d.
General Rules for Investigations into...	0	1

Land Registry:

	s.	d.
General Orders, Directions, and Forms relating to Proceedings on Application for first Registration of Title, and on Transfer and other Dispositions of Land on the Register, &c.	1	6
Land Transfer Act, 1875, Rules and Orders	1	0

Friendly Societies:

	s.	d.
Memorandum for the Guidance of, followed by the principal Provisions of the Acts and Treasury Regulations	0	4
Regulations of Treasury under the Industrial and Provident Societies Act, 1876 ...	0	6

KNIGHT & CO., 90 FLEET STREET.

Probate and Divorce:

CONTENTIOUS.

	s.	d.
Rules and Orders in Contentious Business (1862)...	1	0
Amended and Additional Rules and Orders (1865)	0	2
Additional Rules and Orders (1874) ...	0	3

NON-CONTENTIOUS.

	s.	d.
Rules, Orders, and Instructions for the Registrars of the Principal Registry in Non-Contentious Business (1862)	1	4
Amended Rule and Order in Non-Contentious Business (1865)	0	2
Amended Rules and Orders for the Registrars of the Principal Registry in Non-Contentious Business (1871)	0	1
Rules, Orders, and Instructions for the District Registrars in Non-Contentious Business (1863)	1	4
Amended Rules and Orders for the District Registrars in Non-Contentious Business (1871)	0	1
Table of Fees to be taken in the Principal Registry of the Court of Probate, and in the District Registries thereof (1874)...	1	0
Tables of Fees to be taken by Proctors, &c., practising in the Principal and District Registries of the Court of Probate in Non-Contentious Business (1874)	0	6

Probate Parchment Forms ...
Double Probate ditto ...
Administration ditto ...
Administration de bonis non Forms ...
Ditto with Will annexed Forms ...
Now Supplied by Inland Revenue.

	s.	d.
Court of Probate, Debtors' Act, 1869, Rules for Regulating the practice under, and carrying into effect the first part of the said Act ...	0	2

Chancery:

	s.	d.
Chancery Fund Rules, Chancery Fund Orders, and Chancery Funds (Lunacy) Orders, 1872 ...	0	6
Chancery Funds Consolidated Order...	0	3

Divorce:

	s.	d.
Rules and Regulations for Her Majesty's Court of Divorce, &c.	1	0
Additional Rules and Regulations for ditto	0	2
Table of Fees to be taken in ditto ...	0	3
Debtors' Act, 1869, Rules for regulating the Practice under, and carrying into effect the first part of the said Act	0	2
Amended and Additional Rules, 1875	0	2
Additional Rules and Regulations, 1877	0	2

KNIGHT & CO., 90 FLEET STREET.

Employers and Workmen Act, 1875:

	s.	d.
Rules for Court of Summary Jurisdiction ...	0	2
Forms under same, see separate List.		

Supreme Court:

		s.	d.
Rules of (Costs), being the additional Rules made by Order in Council of 12th August, 1875 ...	...	0	6
Ditto, December, 1875, and February, 1876 ...	each	0	2
Order as to Fees and Percentages, 22nd April, 1876	...	0	3
Rules of—June, 1876	each	0	2
December, 1876; May, 1877; June, 1877 ...	,,	0	1
Lord Chancellor's Order, 19th June, 1877	...	0	1

Trade Marks:

		s.	d.
Up to August 7, 144 Nos.			
Index—namely, May to December, 1876; January to June, 1877; June to December, 1877... ...	each	3	0
Trade Marks Journal, published in parts (parts 1 to 134 already published)	each	1	0
Trade Marks Registration Act, 1875, Rules under	...	1	0
Forms under ditto, A, B, C, D	per doz.	0	6

Cruelty to Animals Act, Forms under—viz.:

		s.	d.
Application for License		0	0½
Certificates A to F	each	0	0½

Merchant Shipping Acts, 1854 to 1876:

	s.	d.
General Rules for formal Investigations into Shipping Casualties, 1876	0	1
General Rules for Courts of Survey in the United Kingdom, 1876	0	3

Hackney Carriages and Metropolitan Stage Carriages:

	s.	d.
Abstract of Laws in force in Metropolitan District (1871 Ed.)	0	2
Distances (1874 Ed.)	1	0

KNIGHT & CO., 90 FLEET STREET.

www.ingramcontent.com/pod-product-compliance
Lightning Source LLC
Chambersburg PA
CBHW031352230426
43670CB00006B/520